"Oh Excellent Air Bag"

CURSE

The original text of the introduction is copyright © 2016 by Mike Jay. The remaining texts are believed to be in the public domain and free from additional rights unless otherwise noted. Usual copyright applies to the design and all other aspects of this publication.

First published in 2016 by
PDR Press
The Public Domain Review
Open Knowledge Foundation
St John's Innovation Centre
Cowley Road
Cambridge CB4 0WS
United Kingdom

pdrpress@publicdomainreview.org
www.publicdomainreview.org

EDITED BY
Adam Green

DESIGNED BY
Nicholas Jeeves
www.nicholasjeeves.com

TYPOGRAPHY
Body text set in 10.5pt on 14.5pt Adobe Caslon Pro, designed in 1990 by Carol Twombly, based on William Caslon's own specimen pages printed between 1734 and 1770

A CIP catalogue record for this book is available from the British Library.

10 9 8 7 6 5 4 3 2 1

"Oh Excellent Air Bag"
Under the Influence of Nitrous Oxide
1799–1920

Contents

9
A note from the editor

12
Introduction by Mike Jay

24
"I began to feel a slight glow in the cheeks"
HUMPHRY DAVY, 1799

32
"As if composed of finely vibrating strings"
THOMAS BEDDOES, 1799

36
"I did burst into a violent fit of laughter
and capered about the room"
P.M. ROGET, S.T. COLERIDGE ET AL, 1799

53
"Oh excellent air bag"
ROBERT SOUTHEY, 1799

55
"As if by the wand of a wizard entranc'd"
RICHARD POLWHELE, 1800

67
"Inflated with supreme intensity"
CHRISTOPHER CAUSTIC, 1803

72
"Blown by a rudely malicious blast
into a world of reptiles"
WILLIAM BARTON, 1808

82
"Dancing, jumping, kicking, fencing,
and occasionally boxing"
MOSES THOMAS, 1814

87
"This is a queer world"
BENJAMIN BLOOD, 1874

92
"Good and evil reconciled in a laugh!"
WILLIAM JAMES, 1882

98
"Om! Om! Om! Om! Om! Om! Om!"
THEODORE DREISER, 1914

117
"I took my station on a base of
infinite nothingness"
ANONYMOUS, 1920

124
Index of Exclamations and Similes

127
Gallery of Images

A cartoon featured in William Barton's 1808 doctoral thesis on the "chemical and exhilarating effects of nitrous oxide gas".

A note from the editor

Before the gas takes hold, a little word about the texts and some decisions taken regarding their presentation. The diligent reader will notice some perhaps unfamiliar spellings ("head-ach", "chearful", "extacy", etc.), orthographic quirks we've decided to carry over from the original texts, and like-wise with certain inconsistencies across the different texts ("air bag" and "air-bag", for example, and a variety of punctuation styles). This has come from a desire to keep as true to the originals as possible. Despite this instinct, we have, however, taken a few liberties with an eye to making the reading experience as enjoyable and coherent as possible. Although all the footnotes you see are from the originals themselves (there are no intrusions from the editor in this respect), a few have been shortened, and some removed entirely. This is most obviously the case in the Moses Thomas piece, as he seemed to use the device to carry on a wholly irrelevant conversation to that going on in the main text, and we felt it was distracting to say the least. So we removed the lot. Likewise for "The Pneumatic Revellers" and *Terrible Tractoration!!*, where the extensive footnotes either did not add much or were simply repeating material found elsewhere in the volume. Due to concerns about clarity, we gave the layout of the Theodore Dreiser play, as it appeared in its original 1916 publication, a complete revamp. For similar reasons we've edited slightly the collection of accounts of Davy's circle, removing the sign-offs, and altering the format of the sub-headers. Lastly, a little word about the dates stated in the contents page. Where the text describes an experience of inhaling nitrous oxide, the date refers not to the date of this account's publication, but rather to the year in which the experience was had (with reversion to the earliest date, should there be multiple). Where a specific experience is not detailed, the date refers to the year of composition.

Here follows a list of the editions used, in order of appearance:

DAVY, HUMPHRY. *Researches, Chemical and Philosophical: Chiefly Concerning Nitrous Oxide, or Dephlogisticated Nitrous Air, and its Respiration.* London, 1800.

BEDDOES, THOMAS. *Notice of Some Observations Made at the Medical Pneumatic Institution.* Bristol, 1799.

SOUTHEY, ROBERT. A letter to his brother Thomas Southey, 12 July 1799. MS: British Library, Add MS 30927. ALS; 4p.

DAVY, JOHN, ed. *Fragmentary Remains, Literary and Scientific, of Sir Humphry Davy, Bart.* London, 1858.

POLWHELE, RICHARD. "The Pneumatic Revellers", *Anti-Jacobin Review and Magazine, or, Monthly Political and Literary Censor* VI (1800): 109–118.

CAUSTIC, CHRISTOPHER. *Terrible Tractoration!! A Poetical Petition Against Galvanising Trumpery, and the Perkinistic Institution.* 2nd edition. London, 1803.

BARTON, WILLIAM P. C. *Dissertation on the Chymical Properties and Exhilarating Effects of Nitrous Oxide Gas, and its Application to Pneumatick Medicine.* Philadelphia, 1808.

THOMAS, MOSES. *A Cursory Glimpse of the State of the Nation, on the Twenty-Second of February, 1814, Being the Eighty-First Anniversary of the Birth of Washington, or, A Physico-Politico-Theologico Lucubration Upon the Wonderful Properties of Nitrous Oxide, or the Newly Discovered Exhilarating Gas, in its Effects upon the Human Mind, and Body.* Philadelphia, 1814.

BLOOD, BENJAMIN PAUL. *The Anaesthetic Revelation and the Gist of Philosophy*. New York, 1874.

JAMES, WILLIAM. "On Some Hegelisms", *MIND* VII.26 (1882): 186–208.

DREISER, THEODORE. "Laughing Gas", in *Plays of the Natural and Supernatural*. New York, 1916.

ANONYMOUS. "The Chair of Metaphysics", *The Atlantic Monthly* XXIII (1920): 421–424.

Many thanks to Benjamin Myers for his help in getting the relevant words from the above books into this one, and to Lauren Washuk for her proof-reading prowess. Thanks also to Mike Jay, not only for his wonderful introduction, but also for his very helpful suggestions regarding texts to include. And finally, as ever, thanks to Nicholas Jeeves for all the hours he has put into making the book happen, not just from a design point-of-view, but also in relation to his ever-valuable and wise editorial counsel.

Adam Green,
London, May 2016

Introduction

MIKE JAY

On Boxing Day of 1799, the twenty-year-old chemist Humphry Davy — later to become Sir Humphry, inventor of the miners' lamp, President of the Royal Society, and domineering genius of British science — stripped to the waist, placed a thermometer under his armpit, and stepped into a sealed box specially designed by the engineer James Watt for the inhalation of gases. As he did so, he requested the physician Robert Kinglake to release twenty quarts of nitrous oxide into it every five minutes for as long as he could retain consciousness. After an hour and a quarter, by which time he estimated that his system was saturated, Davy stepped unsteadily out of the box and proceeded to inhale a further twenty quarts of the gas from a series of silk bags. His description of what followed (p.26) would become a public sensation and a symbol of the heroic commitment to science that would define the coming century.

Davy's experiment was the climax of a freewheeling programme of consciousness expansion into which he had already co-opted some of the most remarkable figures of his day, and it would inspire a century and more of first-person descriptions that combined the scientific, the poetic, and the philosophical in previously unimagined ways. This volume presents a selection of the most striking and celebrated examples, some reproduced here for the first time. With the rediscovery of nitrous oxide by the drug culture of the twenty-first century, these can be appreciated not only as a remarkable body of scientific reportage but — long before the term was coined — one of the great flowerings of psychedelic literature.

The story begins in the laboratory of the Pneumatic Institution in Hotwells, a run-down spa at the foot of the Avon Gorge outside Bristol, where the young Humphry Davy had been taken on as laboratory assistant. Originally developed as a centre for medical therapies to rival nearby Bath, by 1799 Hotwells had dwindled to a downmarket cluster of cheap clinics and miracle-cure outfits offering hydrotherapy or mesmerism to those in the desperate last stages of consumption. The Pneumatic Institution was a new arrival with revolutionary ambitions. Its founder, the brilliant, maverick doctor Thomas Beddoes, believed that the new gases with which he and his assistant were experimenting had the power to put the treatment of this most lethal of diseases onto a proper scientific footing for the first time, and in the process to transform the art of medicine.

Davy inhaled his first lungful of nitrous oxide in April 1799, as he and Beddoes were working their way systematically through the synthesis of gases that might have therapeutic potential. He had set up a chemical reaction: nitrate of ammoniac bubbled in a heated retort, and the escaping gas was collected in a hydraulic bellows before seeping through water into a reservoir tank. Decanting it into an oiled green silk air-bag that the engineer James Watt had designed for dispensing gases to patients, Davy inhaled and was astonished to notice "a highly pleasurable thrilling in the chest and extremities". Filling his lungs repeatedly, he felt "an irresistible propensity to action" which he indulged by "shouting, leaping and running" around the laboratory in ecstasy.

Within days Davy began offering the gas to his friends and soliciting their descriptions of its effects. The first guinea-pig was the young poet Robert Southey, whose reaction was as effusive as Davy's own: "the atmosphere of the highest of all possible heavens must be composed of this gas!" Southey's ecstatic report to his brother Tom (p.53) set the tone for the explorations that were to follow.

In the early summer of 1799 the trials of nitrous oxide began in earnest. During the evenings, when surgery hours at the Pneumatic Institution were over, the nitrate of ammoniac reaction would be set bubbling in the upstairs drawing room and the green silk bag would fill for Davy and Beddoes' circle of local doctors and their patients, chemists, playwrights, surgeons, lecturers, and poets. Davy was master of ceremonies and led from the front — by his own account, inhaling the gas himself three or four times a day. The laboratory became a philosophical theatre in which the boundaries between experimenter and subject, spectator and performer were blurred to fascinating effect, and the experiment took on a life of its own.

Although the trials commenced within a medical framework, they came to focus increasingly on questions of metaphysics and, in particular, language. Davy was struck by the poverty of the "language of feeling" available to his subjects, and the awkwardness of their attempts to put their experiences into words. The standard medical question "how do you feel?" took on imponderable, existential dimensions. The subjects were not mentally impaired by the gas, but the sensations and insights it produced were somehow beyond the reach of words. As Davy himself put it, "I have sometimes experienced from nitrous oxide, sensations similar to no others, and they have consequently been indescribable". James Thompson, one of the volunteers, captured the magnitude of the task precisely: "We must either invent new terms to express these new and peculiar sensations, or attach new ideas to old ones, before we can communicate intelligibly with each other on the operation of this extraordinary gas."

Davy instituted a loose reporting protocol, asking every volunteer to produce a short written description of their experience. He also took enthusiastically to experimenting on his own. On full moon nights in particular, he would wander down the Avon Gorge with a bulging green silk bag and notebook, inhaling the gas under the stars and scribbling snatches of poetry and philosophical insights. On one occasion he made himself conspicuous by passing out and, on recovery, was obliged to "make a bystander acquainted with the pleasure I experienced by laughing and stomping". He noted an element of compulsion in his use, confessing that "the desire to breathe the gas is awakened in me by the sight of a person breathing, or even by that of an air-bag or air-holder". He began to push his experiments into more dangerous territory. He tried the gas in combination with different stimulants, drinking a bottle of wine methodically in eight minutes flat and then inhaling so much gas he passed out for two hours. He experimented with nitric oxide, which turned to nitric acid in his mouth, burning his tongue and palate, and with "hydrocarbonate" — hydrogen and carbon dioxide — which left him comatose, the air-bag fortunately falling from his lips. On recovering, he "faintly articulated: 'I do not think I shall die'".

By the end of the summer, the energy of the trials was dissipating: for most of the volunteers the novelty wore off after a few sessions. Davy's experiments became increasingly solitary, partially focused on resolving technical questions such as how much gas was absorbed into the bloodstream and whether it should be classified as a stimulant or a sedative, but also searching for a framework — scientific, poetic, or philosophical — to account for its effects. In this he was assisted by the arrival of Samuel Taylor Coleridge, who returned to Bristol in October from an extended visit to Germany.

Coleridge and Davy's friendship would evolve and endure through the many phases of their future careers, but it began with a green silk bag. Coleridge was captivated by the young chemist: "Every subject in Davy's mind", he wrote, "has the principle of vitality. Living thoughts spring up like turf under his feet". Davy was equally swept up in his new friend's expansive vision, which he felt had the power to transform science as much as poetry. Coleridge had returned from Germany in thrall to the new idealistic turn in its philosophy: the theories of Immanuel Kant and the emerging "Naturphilosophie", according to which the human mind was the ultimate source of our reality, and the material world merely an illusion projected by it. The dissociative effects of nitrous oxide, during which consciousness seemed to exist in a dimension beyond the physical body, made compelling sense of this insight. The conclusion of Davy's Boxing Day experiment — "nothing exists but thoughts" — would echo repeatedly through the experiments that followed.

After his climactic Boxing Day experiment, Davy began writing at top speed. By Easter of 1800, he had produced a 580-page monograph on the new gas and its effects. *Researches, Chemical and Philosophical: Chiefly Concerning Nitrous Oxide, or Dephlogisticated Nitrous Air, and its*

Respiration described the synthesis of the gas, its effect on animals and animal tissue, and, in an unprecedented final section, the descriptions of the effects of nitrous oxide intoxication on himself and two dozen further subjects, including Coleridge and Southey (p.36). The report combined two mutually unintelligible languages — organic chemistry and subjective description — to create a groundbreaking hybrid, a poetic science that could encompass both the chemical causes of the experience and its philosophical consequences.

Even before the *Researches* were published, however, the backlash began. In May 1800, the *Anti-Jacobin Review*, a conservative journal that had long had Beddoes in its sights as a French revolutionary sympathiser, lampooned his early reports of the Pneumatic Institution's discovery in verse. Beddoes' announcement of a revolution in the human condition set the scene for a mock-symposium in which the likes of Robert Southey declaimed their gaseous revelations in pompous doggerel before dissolving into inane laughter and incoherence (p.55). Another verse satire, *Terrible Tractoration!!*, written three years later as part of a wider assault on the medical world, offers a further example of the irresistible temptation to parody that the nitrous oxide experience presented: a lungful of gas and a moment of soaring cosmic revelation, followed by the flatulent deflation of the air-bag together with the philosopher's absurd pretensions (p.67).

Despite such public mockery, Davy's experiments were promptly reproduced by the network of chemists who had been following his progress through the bulletins he sent to *Nicholson's Journal*, their unofficial clearing-house. A group of London amateur chemists called the Askeian Society replicated his experiments in March 1800; one of its members recorded that, on inhaling the gas, he "had the idea of being carried violently upwards in a dark cavern with only a few glimmering lights". The first published report from America appeared in 1808: William Barton, a medical student at Philadelphia University, took up Davy's challenge of describing "original sensations" where "no analogous feelings have previously existed", and found himself reaching for the poetry of Shakespeare, Milton, and Rousseau (p.72). By 1814 the experience had made the leap from the laboratory to the theatre: the *Philadelphia Gazette* was carrying small ads for public lectures by a Dr Jones on nitrous oxide and its effects.

The scene was described by the printer and pamphleteer Moses Thomas in an eccentric tract subtitled *A Physico-Politico-Theological Lucubration on the Wonderful Properties of Nitrous Oxide*, which segues from vivid reportage to a disquisition on the question of invading Canada (p.82). Dr Jones was careful to set the stage with sober scientific intent, but it's clear that he was also manipulating his audience to create a diverting public performance. The caged audience created the expectation that the "antic" effects of the gas might break the boundaries of decorum, and the drama of selecting individual audience members encouraged a lively and extroverted response: anyone who, like Coleridge, "remained motionless, in great ecstacy" would risk being booed off stage. Not unlike today's stage hypnotism shows, the entertainment was provided by an audience who already had a firm idea of what such entertainment should look like.

Nitrous oxide rapidly became a staple of lecture halls, variety theatres, fairgrounds, and carnivals. The chemical apparatus required a substantial initial outlay, making it prohibitively expensive to indulge in private but lucrative for entrepreneurs offering it as a novelty to the paying public. The new class of "professional" experimenters, naturally, never tired of stressing the dangers of synthesising or inhaling the gas without expert supervision. By 1824, a nitrous oxide show was part of the variety bill at London's Adelphi theatre, as part of a programme of "Uncommon Illusions, Wonderful Metamorphoses, Experimental Chemistry, Animated Paintings etc.". Humphry Davy's name was prominent on the poster as, at other venues, was the ecstatic testimony of Robert Southey: names to conjure with, the former being now President of the Royal Society and the latter Poet Laureate. Also prominent on the Adelphi poster was a new epithet for nitrous oxide: "The Laughing Gas".

The German chemist Christian Schoenbein, who would later invent the fuel cell and discover ozone, was one of many who witnessed the

Adelphi performances. He recorded that the scientific preamble was interrupted by a heckler (a plant?) from the audience who shouted that the wonders of the gas were "all nonsense and humbug"; of course, the heckler was the first to be invited up on stage, where, after receiving his dose, he "beat around him like a madman and assaulted the 'Experimentator'". Schoenbein was wonderfully entertained, and convinced that the gas was destined to become still more popular: "Maybe it will become the custom for us to inhale laughing gas at the end of a dinner party, instead of drinking champagne".

Such shows were ideally suited to the travelling carnival circuit in America, where the novelty could be repeated in a different town night after night for years. One of the first itinerants to make a success of them was an eighteen-year-old Samuel Colt, later to design the first mass-produced revolver, who toured a nitrous show around the East Coast, from Canada to Maryland, in a tent emblazoned with Robert Southey's endorsement that it must be "the atmosphere of the highest of all possible heavens" (p. 53).

It was through the American "laughing gas shows" that nitrous oxide finally found its defining medical application. During the 1840s, a travelling temperance campaigner named Gardiner Quincy Colton was plying carnivals and show-grounds with an exhibit he called *Court of Death*: a huge diorama painting depicting the evils of drink and the pits of hell, the backdrop for a fire-and-brimstone lecture. At twenty-five cents a time, the same as a magic lantern show or fairground ride, he found himself struggling for business until he had the idea of adding nitrous oxide to the mix. He adopted the now-familiar trappings: twelve strong men to protect the performers from unpredictable reactions, "safe" dress seats for the ladies, and a gentlemen-only rule for the volunteers. Colton proved an expert master of ceremonies, adept at interpreting the effects of the gas to reinforce the moral of his show. Nitrous oxide, he explained, had the uncanny effect of materialising original sin. Safely, temporarily, and reversibly, it would expose the inner natures of the audience, revealing how bestial they might become if they failed to make temperance their guiding light.

In 1844, Colton took his gas-fuelled *Court of Death* show to Hartford, Connecticut, where a young dentist called Horace Wells happened to be in the audience. One of the volunteer performers was gripped by the familiar "antic" mania under the influence and rushed into one of the front benches, smashing his shin against it. He carried on entirely unaware of his injury until the gas wore off, at which point he registered severe pain. Wells examined him and, finding the injury to be quite serious, was struck by the idea that nitrous oxide might make an effective dental anaesthetic. He himself had a wisdom tooth in need of extraction and had been reluctant to entrust it to any of his colleagues; he asked Colton if he would be prepared to administer the gas to him during the operation. Colton agreed, and Wells' tooth was extracted painlessly. The experience was a revelation for Wells, both professional and personal. He emerged, not unlike Davy, proclaiming his vision in prophetic manner: "A new era in tooth-pulling!"

Colton abandoned *Court of Death* and reinvented himself as the Colton Dental Association, offering nitrous anaesthesia to dental patients. The "painless method" was spread by word of mouth for nearly twenty years before professional dentists were prepared to accept it. It was widely adopted only after Colton set up in New York in 1863, by which time he had already franchised his association across several American cities and used the gas in at least 75,000 surgical extractions.

Nitrous oxide was finally hailed as a miracle of modern medicine, just as Beddoes and Davy had predicted, but its strange and ineffable dimensions were not easily dismissed. One of the thousands who passed through the Colton Dental Association in the 1860s was Benjamin Paul Blood, a farmer, bodybuilder, calculating prodigy, and tireless pamphleteer who first experienced the effects of the gas during dental surgery in upstate New York. As many had before him, he felt a powerful conviction that the secret of life had been briefly and tantalisingly laid bare under the influence; unlike most, he proceeded to synthesise the gas himself and repeat the experience. In 1874, after ten years of self-experimentation, he produced a pamphlet on the subject entitled *The Anaesthetic Revelation and the Gist of Philosophy* (p.87).

Self-published and sent unannounced to everyone Blood could think of, *The Anaesthetic Revelation* led to the formation of a loose circle of correspondence. It found readers among the ranks of the recently-formed Society for Psychical Research in Britain: the spritualist Edmund Gurney, the Nobel-winning physicist William Ramsay, the leading British psychiatrist Sir James Crichton-Browne, and the art historian John Addington Symonds were among the self-experimenters who contributed their observations on nitrous oxide to the SPR journals. But Blood's most influential convert was William James, professor of psychology at Harvard and founder of the American Society for Psychical Research, who encountered his pamphlet in the only form in which it was publicly reproduced, an anonymous digest in *The Atlantic Monthly*.

James recognised immediately that the experience it described might prove relevant to what he had come to regard as a dilemma in modern thought. As a physician and psychologist he was convinced that scientific materialism provided the answers to many questions that, prior to its development, could not even properly be asked. Yet he felt that its method was intrinsically loaded against certain classes of mental phenomena, particularly those states and experiences usually referred to as "mystical" or "religious". These had become pejorative terms within science, synonymous with "unverifiable", "subjective", even "meaningless"; but to James, this was a failing not of religion but of science. If science was to be the fundamental account of reality, it needed to find a way to make sense of them.

James was attracted to the philosophical idealism of Georg Hegel and its claim that truth could — indeed, could only — be arrived at by accepting both a proposition and its opposite. Thus he was particularly struck by the way in which Blood framed his discovery in contradictory terms, referring to the nitrous-induced state, for example, as a "condition (or uncondition)". Setting up the now-familiar reaction of ammonium nitrate with retorts, tubes, and cloth bags, James embarked on a practical solution to his conundrum. No less than for Blood, his experience was a revelation. As it had for Davy, the dissociative properties of the gas and the radical abstractions of German idealism proved a magnificently fertile combination, generating a transcendent state in which all dualities and oppositions melted into a euphoric epiphany that made sense of everything (p.92).

James expanded on this revelation in his most enduring work, *The Varieties of Religious Experience* (1902), which coalesced around the synthesis that nitrous oxide revealed to him: the phenomena of "extraordinary consciousness", whatever the truth or otherwise of their revelations, are "psychologically real". In what has become one of the book's most frequently-quoted passages, James attributes this realisation explicitly to his inhalation of nitrous oxide:

> One conclusion was forced upon my mind at that time, and my impression of its truth has ever since remained unshaken. It is that our normal waking consciousness, rational consciousness as we call it, is but one special type of consciousness, whilst all about it, parted from it by the filmiest of screens, there lie potential forms of consciousness entirely different... No account of the universe in its totality can be final which leaves these other forms of consciousness quite disregarded.

Nitrous oxide entered the twentieth century firmly defined as a surgical and dental anaesthetic, but its mysterious effects on consciousness lingered in the public imagination. In 1914, Theodore Dreiser wrote a one-act play entitled "Laughing Gas" in which a doctor has a supernatural experience while undergoing surgery, perhaps the only dramatic work in which "The Rhythm of the Universe" is allocated a speaking part (p.98). Both in fiction and in reality, the prosaic setting of the dentist's chair emerged as a seat of unexpected revelations, tantalisingly hard to capture in words but imbued with the conviction of a perennial philosophy beyond time and language. In 1920, *The Atlantic Monthly*, in which Benjamin Blood and William James' revelations had previously appeared, printed the anonymous report of an astonished patient whose routine dental treatment became a voyage into the deepest secrets of the universe (p.117). In 1936, P.G. Wodehouse published his novel *Laughing Gas*, in which nitrous oxide is the source of a fourth-dimensional mix-up in the dentist's chair that creates a mind-swap between two anaesthetised characters.

It was through medical channels that nitrous oxide first made its entry into modern drug culture. During the 1960s, especially in the United States, tanks of gas were occasionally "liberated" from their hospital duty: the Grateful Dead took to carrying one in their tour bus, and Hollywood poolside parties were enlivened by filling lilos and inflatable toys from them. But surgical tanks and inhalation masks made it all too easy for recreational users to pass out and asphyxiate under the influence, and a few well-publicised deaths were enough to halt the craze in its tracks. Its recent popularity exploits a different source, the miniature bulbs of compressed gas produced for whipping cream which can be emptied into balloons for inhalation. These are more accessible, cheaper, and also safer: if the user loses consciousness the balloon simply falls from the lips and allows normal breathing to resume.

The first sign of this discovery could be discerned at summer music festivals, where dawn would reveal a carpet of small silver gas cannisters and dew-soaked balloons scattered across the ground. Today these telltale signs can be found discarded in any urban park or street corner. The famously indescribable, unaccountably hilarious, and oddly profound experience for so long mediated by chemists, dentists, or surgeons is now familiar to millions. Though its devotees still include scientific truth-seekers and mystics in the tradition of Humphry Davy or William James, the majority are perhaps closer in spirit to the crowds at the nineteenth-century laughing gas shows, briefly stepping out of normal consciousness into an effervescent blast of pleasure. The current scene shows little sign of generating a literature as rich as that collected here; perhaps the old problem of capturing the experience in words has been resolved with new languages such as that of electronic music and psychedelic art. But the balloon, ubiquitous technology and symbol of the modern nitrous experience, will always pay unconscious tribute to the Pneumatic Institution's green silk bag.

*

MIKE JAY *has written extensively on scientific and medical history and is a specialist in the study of drugs. His books include* The Atmosphere of Heaven: The Unnatural Experiments of Dr Beddoes and His Sons of Genius *(2009),* High Society: Mind-Altering Drugs in History and Culture *(2012), and* Emperors of Dreams: Drugs in the Nineteenth Century *(2000).*

1

"I began to feel a slight glow in the cheeks"

Extracts from HUMPHRY DAVY's *Researches, Chemical and Philosophical* (1800) in which he recounts his early self-experiments with nitrous oxide, from April 1799 through to June 1800. Nitrous oxide had been first synthesised by Joseph Priestley 27 years earlier, but Davy was the first to fully investigate its effects when inhaled — effects of which he became perhaps a little too fond.

On April 11th, I made the first inspiration of pure nitrous oxide; it passed through the bronchia without stimulating the glottis, and produced no uneasy feeling in the lungs. The result of this experiment, proved that the gas was respirable, and induced me to believe that a farther trial of its effects might be made without danger. On April 16th, Dr. Kinglake being accidentally present, I breathed three quarts of nitrous oxide from and into a silk bag for more than half a minute, without previously closing my nose or exhausting my lungs.

The first inspirations occasioned a slight degree of giddiness. This was succeeded by an uncommon sense of fulness of the head, accompanied with loss of distinct sensation and voluntary power, a feeling analogous to that produced in the first stage of intoxication; but unattended by pleasurable sensation. Dr. Kinglake, who felt my pulse, informed me that it was rendered quicker and fuller.

This trial did not satisfy me with regard to its powers; comparing it with the former ones I was unable to determine whether the operation was stimulant or depressing.

I communicated the result to Dr. Beddoes, and on April the 17th, he was present, when the following experiment was made. Having previously closed my nostrils and exhausted my lungs, I breathed four quarts of nitrous oxide from and into a silk bag. The first feelings were similar to those produced in the last experiment; but in less than half a minute, the respiration being continued, they diminished gradually, and were succeeded by a sensation analogous to gentle pressure on all the muscles, attended by an highly pleasurable thrilling, particularly in the chest and the extremities. The objects around me became dazzling and my hearing more acute. Towards the last inspirations, the thrilling increased, the sense of muscular power became greater, and at last an irresistible propensity to action was indulged in; I recollect but indistinctly what followed; I know that my motions were various and violent.

Between May and July, I habitually breathed the gas, occasionally three or four times a day for a week together; at other periods, four or five times a week only. The doses were generally from six to nine quarts; their effects appeared undiminished by habit, and were hardly ever exactly similar. Sometimes I had the feelings of intense intoxication, attended with but little pleasure; at other times, sublime emotions connected with highly vivid ideas; my pulse was generally increased in fulness, but rarely in velocity.

Between September and the end of October, I made but few experiments on respiration, almost the whole of my time being devoted to chemical experiments on the production and analysis of nitrous oxide.

At this period my health being somewhat injured by the constant labour of experimenting, and the perpetual inhalation of the acid vapours of the laboratory, I went into Cornwal; where new associations of ideas and feelings, common exercise, a pure atmosphere, luxurious diet and moderate indulgence in wine, in a month restored me to health and vigor.

Nov. 27th. Immediately after my return, being fatigued by a long journey, I respired nine quarts of nitrous oxide, having been precisely thirty-three days without breathing any. The feelings were different from those I had experienced in former experiments. After the first six or seven inspirations, I gradually began to lose the perception of external things, and a vivid and intense recollection of some former experiments passed through my mind, so that I called out "what an amazing concatenation of ideas".

On December 26th, I was inclosed in an air-tight breathing-box,[1] of the capacity of about 9 cubic feet and half, in the presence of Dr. Kinglake.

After I had taken a situation in which I could by means of a curved thermometer inserted under the arm, and a stop-watch, ascertain the alterations in my pulse and animal heat, 20 quarts of nitrous oxide were thrown into the box.

---

1. The plan of this box was communicated by Mr. Watt. An account of it will be detailed in the *Researches*.

For three minutes I experienced no alteration in my sensations, though immediately after the introduction of the nitrous oxide the smell and taste of it were very evident.[2]

In four minutes I began to feel a slight glow in the cheeks, and a generally diffused warmth over the chest, though the temperature of the box was not quite 50°. I had neglected to feel my pulse before I went in; at this time it was 104 and hard, the animal heat was 98°. In ten minutes the animal heat was near 99°, in a quarter of an hour 99.5°, when the pulse was 102, and fuller than before.

At this period 20 quarts more of nitrous oxide were thrown into the box, and well-mingled with the mass of air by agitation.

In 25 minutes the animal heat was 100°, pulse 124. In 30 minutes, 20 quarts more of gas were introduced.

My sensations were now pleasant; I had a generally diffused warmth without the slightest moisture of the skin, a sense of exhilaration similiar to that produced by a small dose of wine, and a disposition to muscular motion and to merriment.

In three quarters of an hour the pulse was 104, and animal heat not 99.5°, the temperature of the chamber was 64° The pleasurable feelings continued to increase, the pulse became fuller and slower, till in about an hour it was 88°, when the animal heat was 99°.

20 quarts more of air were admitted. I had now a great disposition to laugh, luminous points seemed frequently to pass before my eyes, my hearing was certainly more acute and I felt a pleasant lightness and power of exertion in my muscles. In a short time the symptoms became stationary; breathing was rather oppressed, and on account of the great desire of action, rest was painful.

I now came out of the box, having been in precisely an hour and a quarter.

The moment after, I began to respire 20 quarts of unmingled nitrous oxide. A thrilling extending from the chest to the extremities

---

2. The nitrous oxide was too diluted to act much; it was mingled with near 32 times its bulk of atmospheric air.

was almost immediately produced. I felt a sense of tangible extension highly pleasurable in every limb; my visible impressions were dazzling and apparently magnified, I heard distinctly every sound in the room and was perfectly aware of my situation.[3] By degrees as the pleasurable sensations increased, I lost all connection with external things; trains of vivid visible images rapidly passed through my mind and were connected with words in such a manner, as to produce perceptions perfectly novel. I existed in a world of newly connected and newly modified ideas. I theorised; I imagined that I made discoveries. When I was awakened from this semi-delirious trance by Dr. Kinglake, who took the bag from my mouth, Indignation and pride were the first feelings produced by the sight of the persons about me. My emotions were enthusiastic and sublime; and for a minute I walked round the room perfectly regardless of what was said to me. As I recovered my former state of mind, I felt an inclination to communicate the discoveries I had made during the experiment. I endeavoured to recall the ideas, they were feeble and indistinct; one collection of terms, however, presented itself: and with the most intense belief and prophetic manner, I exclaimed to Dr. Kinglake, "Nothing exists but thoughts!—the universe is composed of impressions, ideas, pleasures and pains!"

About three minutes and half only, had elapsed during this experiment, though the time as measured by the relative vividness of the recollected ideas, appeared to me much longer.

Not more than half of the nitrous oxide was consumed. After a minute, before the thrilling of the extremities had disappeared, I breathed the remainder. Similar sensations were again produced; I was quickly thrown into the pleasurable trance, and continued in it longer than before. For many minutes after the experiment, I experienced the thrilling in the extremities, the exhilaration continued nearly two hours. For a much longer time I experienced the mild enjoyment before described connected with indolence; no depression or feebleness followed. I ate my dinner with great appetite and found myself

---

3. In all these experiments after the first minute, my cheeks became purple.

lively and disposed to action immediately after. I passed the evening in executing experiments. At night I found myself unusually cheerful and active; and the hours between eleven and two, were spent in copying the foregoing detail from the common-place book and in arranging the experiments. In bed I enjoyed profound repose. When I awoke in the morning, it was with consciousness of pleasurable existence, and this consciousness more or less continued through the day.

Since December, I have very often breathed nitrous oxide. My susceptibility to its power is rather increased than diminished. I find six quarts a full dose, and I am rarely able to respire it in any quantity for more than two minutes and half.

The mode of its operation is somewhat altered. It is indeed very different at different times.

I am scarcely ever excited into violent muscular action, the emotions are generally much less intense and sublime than in the former experiments, and not often connected with thrilling in the extremities.

When troubled with indigestion, I have been two or three times unpleasantly affected after the excitement of the gas. Cardialgia, eructations and unpleasant fulness of the head were produced.

I have often felt very great pleasure when breathing it alone, in darkness and silence, occupied only by ideal existence. In two or three instances when I have breathed it amidst noise, the sense of hearing has been painfully affected even by moderate intensity of sound. The light of the sun has sometimes been disagreeably dazzling. I have once or twice felt an uneasy sense of tension in the cheeks and transient pains in the teeth.

Whenever I have breathed the gas after excitement from moral or physical causes, the delight has been often intense and sublime.

On May 5th, at night, after walking for an hour amidst the scenery of the Avon, at this period rendered exquisitely beautiful by bright moonshine; my mind being in a state of agreeable feeling, I respired six quarts of newly prepared nitrous oxide.

The thrilling was very rapidly produced. The objects around me were perfectly distinct, and the light of the candle not as usual dazzling. The pleasurable sensation was at first local and perceived in the lips and about the cheeks. It gradually however, diffused itself over the whole body, and in the middle of the experiment was for a moment so intense and pure as to absorb existence. At this moment, and not before, I lost consciousness; it was however, quickly restored, and I endeavoured to make a by-stander acquainted with the pleasure I experienced by laughing and stamping. I had no vivid ideas. The thrilling and the pleasurable feeling continued for many minutes; I felt two hours afterwards, a slight recurrence of them, in the intermediate state between sleeping and waking; and I had during the whole of the night, vivid and agreeable dreams. I awoke in the morning with the feeling of restless energy, or that desire of action connected with no definite object, which I had often experienced in the course of experiments in 1799.

I have two or three times since respired nitrous oxide under similar circumstances; but never with equal pleasure.

During the last fortnight, I have breathed it very often; the effects have been powerful and the sensations uncommon; but pleasurable only in a slight degree.

I ought to have observed that a desire to breathe the gas is always awakened in me by the sight of a person breathing, or even by that of an air-bag or an air-holder.

I have this day, June 5th, respired four large doses of gas. The first two taken in the morning acted very powerfully; but produced no thrilling or other pleasurable feelings. The effects of the third breathed immediately after a hearty dinner were pleasant, but neither intense or intoxicating. The fourth was respired at night in darkness and silence after the occurrence of a circumstance which had produced some anxiety. This dose affected me powerfully and pleasantly; a slight thrilling in the extremities was produced; an exhilaration continued for some time, and I have had but little return of uneasiness. 11 P.M.

From the nature of the language of feeling, the preceding detail contains many imperfections; I have endeavoured to give as accurate an account as possible of the strange effects of nitrous oxide, by making use of terms standing for the most similar common feelings.

We are incapable of recollecting pleasures and pains of sense.[4] It is impossible to reason concerning them, except by means of terms which have been associated with them at the moment of their existence, and which are afterwards called up amidst trains of concomitant ideas.

When pleasures and pains are new or connected with new ideas, they can never be intelligibly detailed unless associated during their existence with terms standing for analogous feelings.

I have sometimes experiences from nitrous oxide, sensations similar to no others and they have consequently been indescribable. This has been likewise often the case with other persons. Of two paralytic patients who were asked what they felt after breathing nitrous oxide, the first answered, "I do not know how, but very queer." The second said, "I felt like the sound of a harp." Probably in the one case, no analogous feelings had ever occurred. In the other, the pleasurable thrillings were similar to the sensations produced by music; and hence, they were connected with terms formerly applied to music.

---

4. Physical pleasure and pain generally occur connected with a compound impression, i.e. an organ and some object. When the idea left by the compound impression is called up by being linked accidentally to some other idea or impression, no recurrence, or the slightest possible, of the pleasure or pain in any form will take place. But when the compound impression itself exists without the physical pleasure or pain, it will awaken ideal or intellectual pleasure or pain, i.e. hope or fear. So that physical pleasure and pain are to hope and fear, what impressions are to ideas. For instance, assuming no accidental association, the child does not fear the fire before he is burnt. When he puts his finger to the fire he feels the physical pain of burning, which is connected with a visible compound impression, the fire and his finger. Now when the compound idea of the fire and his finger, left by the compound impression are called up by his mother, saying, "You have burnt your finger," nothing like fear or the pain of burning is connected with it. But when the finger is brought near the fire, i.e. when the compound impression again exists, the ideal pain of burning or the passion of fear is awakened, and it becomes connected with those very actions which removed the finger from the fire.

2

"As if composed of finely vibrating strings"

Extract from THOMAS BEDDOES' *Notice of Some Observations Made at the Medical Pneumatic Institution* (1799); followed by his account reproduced in Humphry Davy's *Researches* (1800). In 1799, Beddoes founded the Pneumatic Institution in Bristol, where he and Davy conducted investigations into various gases, including "dephlogisticated nitrous air".

The author of this notice, notwithstanding the freedom with which he had formerly inhaled oxygen gas, for some time waved the trial of the other. His apoplectic make, joined to the frequent occurrence of a degree of giddiness in others, rendered him timid. The perfect safety, however, with which he saw the two paralytic patients […], and also a third, who is just beginning a course of air, perform the experiment, overcame his scruples. For if the first unpleasant impression, which has been occasionally observed, had the slightest connection with the symptoms of apoplexy, or palsy, he supposed that either these patients, or so other person, would have been injured by the inhalation. He has now taken it daily for some time, in a manner that will be hereafter described.

The first sensations had nothing unpleasant; the succeeding have been agreeable beyond his conception or belief, even after the rapturous descriptions he had heard, and the eagerness to repeat the inhalation which he had so often witnessed. He seems to himself, at the time, (for why should one fear to use ludicrous terms when they are expressive?) to be bathed all over with a bucket full of good humour, and placid feeling pervades his whole frame. The heat of the chest is much greater from a small dose than he ever felt from the largest quantity of oxygen. A constant fine glow, which affects the stomach, led him one day to take an inconvenient portion of food, and to try the air afterwards. It very soon removed the sense of distention, and, he supposes, expedited digestion. He has never tried to bring on the high orgasm; but has generally felt more alacrity at the moment — not one languid, low, *crapulary*, feeling afterwards. It occurred to him that under a certain administration of this gas, sleep might possibly be dispensed with — he is sure that from less sleep he derives more refreshment than for many years past. And his morning alertness equals that of a healthy boy.

During the printing of this paper, the author took a large dose before an excellent judge of the phaenomena of intoxication, who on observing him attentively for some time, exclaimed, *why your eyes twinkle as if you were drunk. You are certainly drunk* — The observation was accurate. Intoxication, as indicated by unsteadiness and stammering at the time, and a random feeling for some hours after, was produced. It subsided into simple high spirits; and no languor followed. Till he took this air, going to a play always brought on a head-ach next morning. Now he rises just as fresh as on other occasions.

Upon the whole, he believes that the Pneumatic Institution might advance a fair claim to the premium, anciently proposed for the discovery of a new pleasure; and he ventures to say that the first slight unpleasant sensations may be obviated by due management, and the gas exhibited with safety even to hysterical females.

Neither my notes nor my recollection supply much in addition to what I formerly stated in the *Notice of Observations at the Pneumatic Institution. Longman.* The gas maintains its first character as well in its effects on me, as in the benefit it confers on some of the paralytic, and the injury it does or threatens to the hysterical and the exquisitely sensible. I find that five or six quarts operate as powerfully as ever. I seem to make a given quantity go farther by holding my breath so that the gas may be absorbed in a great degree without returning into the bag, and therefore, be as little heated before inspiration as possible. — This may be fancy.

After innumerable trials, I have never once felt lassitude or depression. Most commonly I am sensible of a grateful glow *circum præcordia*. This has continued for hours. — In two or three instances only has exhalation failed to be followed by pleasurable feeling, it has never been followed by the contrary. On a few occasions before the gas was exhausted, I have found it impossible to continue breathing.

The pulse at first becomes fuller and stronger. Whenever, after exposure to a cold wind, the warmth of the room has created a glow in the cheeks, the gas has increased this to strong flushing — which common air breathed in the same way, failed to do.

Several times I have found that a cut which had ceased to be painful has smarted afresh, and on taking two doses in succession, the smarting ceased in the interval and returned during the second respiration. I had no previous expectation of the first smarting.

The only time I was near rendering myself insensible to present objects by very carefully breathing several doses in quick succession, I forcibly exclaimed, TONES! — In fact, besides a general thrilling, there seemed to be quick and strong alterations in the degree of illumination of all surrounding objects; and I felt as if composed of finely vibrating strings. On this occasion, the skin seemed in a state of constriction and the lips glued to the mouth-piece, and the mucous membrane of the lungs contracted, but not painfully. However, no contraction or corrugation of the skin could be seen. I am conscious of having made a great number of observations while breathing, which I could never recover.

Immediately afterwards I have often caught myself walking with a hurried step and busy in soliloquy. The condition of general sensation being as while hearing chearful music, or after good news, or a moderate quantity of wine.

3

"I did burst into a violent fit of laughter, and capered about the room"

The section of Davy's *Researches* (1800) which reproduces various accounts of nitrous oxide intoxication, as given by members of Davy's circle to which he introduced the gas in the summer of 1799. Among those attempting to describe the mysterious feelings conjured are DR ROGET of *Roget's Thesaurus* fame; the poets ROBERT SOUTHEY and SAMUEL TAYLOR COLERIDGE; and the heir to the WEDGWOOD pottery empire.

DETAILS of the EFFECTS produced by the RESPIRATION of NITROUS OXIDE upon different INDIVIDUALS furnished by THEMSELVES.

I. DETAIL OF MR. J. W. TOBIN

Having seen the remarkable effects produced on Mr. Davy, by breathing nitrous oxide, the 18th of April; I became desirous of taking some.

A day or two after I breathed 2 quarts of this gas, returning it back again into the same bag, after two or three inspirations, breathing became difficult, and I occasionally admitted common air into my lungs. While the respiration was continued, my sensations became more pleasant. On taking the bag from my mouth, I staggered a little but felt no other effect.

On the second time of making the experiment, I took nearly four quarts, but still found it difficult to continue breathing long, though the air which was left in the bag was far from being impure.

The effects however, in this case, were more striking than in the former. Increased muscular action was accompanied by very pleasurable feelings, and a strong desire to continue the inspiration. On removing the bag from my mouth, I laughed, staggered, and attempted to speak, but stammered exceedingly, and was utterly unable to pronounce some words. My usual state of mind, however, soon returned.

On the 29th, I again breathed four quarts. The pleasant feelings produced at first, urged me to continue the inspiration with great eagerness. These feelings however, went off towards the end of the experiment, and no other effects followed. The gas had probably been breathed too long, as it would not support flame. I then proposed to Mr. Davy, to inhale the air by the mouth from one bag, and to expire it from the nose into another. This method was pursued with less than three quarts, but the effects were so powerful as to oblige me to take in a little common air occasionally. I soon found my nervous system agitated by the highest sensations of pleasure, which are difficult of description; my muscular powers were very much increased, and I went on breathing with great vehemence, not from a difficulty of inspiration, but from an eager avidity for more air. When the bags were exhausted and taken from me, I continued breathing with the same violence, then suddenly starting from the chair, and vociferating with pleasure, I made towards those that were present, as I wished they should participate in my feelings. I struck gently at Mr. Davy and a stranger entering the room at the moment, I made towards him, and gave him several blows, but more in the spirit of good humour than of anger. I then ran through different rooms in the house, and at last returned to the laboratory somewhat more composed; my spirits continued much elevated for some hours after the experiment, and I felt no consequent depression either in the evening or the day following, but slept as soundly as usual.

On the 5th of May, I again attempted to breathe nitrous oxide, but it happened to contain suspended nitrous vapour which rendered it non-respirable.

On the 7th, I inspired 7 quarts of pure gas mingled with an equal quantity of common air, the sensations were pleasant, and my muscular power much increased.

On the 8th, I inspired five quarts without any mixture of common air, but the effects were not equal to those produced the day before; Indeed there were reasons for supposing that the gas was impure.

On the 18th, I breathed nearly six quarts of the pure nitrous oxide. It is not easy to describe my sensations; they were superior to any thing I ever before experienced. My step was firm, and all my muscular powers increased. My senses were more alive to every surrounding impression; I threw myself into several theatrical attitudes, and traversed the laboratory with a quick step; my mind was elevated to a most sublime height. It is giving but a faint idea of the feelings to say, that they resembled those produced by a representation of an heroic scene on the stage, or by reading a sublime passage in poetry when circumstances contribute to awaken the finest sympathies of the soul. In a few minutes the usual state of mind returned. I continued in good spirits for the rest of the day, and slept soundly.

Since the 18th of May, I have very often breathed nitrous oxide.

In the first experiments when pure, its effects were generally similar to those just described.

Lately I have seldom experienced vivid sensations. The pleasure produced by it is slight and tranquil, I rarely feel sublime emotions or increased muscular power.

II. DETAIL OF MR. WM. CLAYFIELD

The first time that I breathed the nitrous oxide, it produced feelings analogous to those of intoxication. I was for some time unconscious of existence, but at no period of the experiment experienced agreeable sensations, a momentary nausea followed it; but unconnected with languor or head-ache.

After this I several times respired the gas, but on account of the fulness in the head and apparent throbbing of the arteries in the brain,[1] always desisted to breathe before the full effects were produced. In two experiments however, when by powerful voluntary efforts I succeeded in breathing a large quantity of gas for some minutes, I had highly pleasurable thrillings in the extremities, and such increase of muscular power, as to be obliged to exert my limbs with violence. After these experiments, no languor or depression followed.

III. LETTER FROM DR. KINGLAKE

In compliance with your desire, I will endeavour to give you a faithful detail of the effects produced on my sensations by the inhalation of nitrous oxide.

My first inspiration of it was limited to four quarts, diluted with an equal quantity of atmospheric air. After a few inspirations, a sense of additional freedom and power (call it energy if you please) agreeably

---

1. In some of these experiments, hearing was rendered more acute.

pervaded the region of the lungs; this was quickly succeeded by an almost delirious but highly pleasurable sensation in the brain, which was soon diffused over the whole frame, imparting to the muscular power at once an encreased disposition and tone for action; but the mental effect of the excitement was such as to absorb in a sort of intoxicating placidity, and delight, volition, or rather the power of voluntary motion. These effects were in a greater or less degree protracted during about five minutes, when the former state returned, with the difference however of feeling more cheerful and alert, for several hours after.

It seemed also to have had the further effect of reviving rheumatic irritations in the shoulder and knee-joints, which had not been previously felt for many months. No perceptible change was induced in the pulse either at or subsequent to the time of inhaling the gas.

The effects produced by a second trial of its powers, were more extensive, and concentrated on the brain. In this instance, nearly six quarts undiluted, were accurately and fully inhaled. As on the former occasion, it immediately proved agreeably respirable, but before the whole quantity was quite exhausted, its agency was exerted so strongly on the brain, as progressively to suspend the senses of seeing, hearing, feeling, and ultimately the power of volition itself. At this period, the pulse was much augmented both in force and frequency; slight convulsive twitches of the muscles of the arms were also induced; no painful sensation, nausea, or languor, however, either preceded, accompanied, or followed this state, nor did a minute elapse before the brain rallied, and resumed its wonted faculties, when a sense of glowing warmth extending over the system, was speedily succeeded by a re-instatement of the equilibrium of health.

The more permanent effects were (as in the first experiment) an invigorated feel of vital power, improved spirits, transient irritations in different parts, but not so characteristically rheumatic as in the former instance.

Among the circumstances most worthy of regard in considering the properties and administration of this powerful aërial agent, may be ranked, the fact of its being (contrary to the prevailing opinion)[2] both highly respirable, and salutary, that it impresses the brain and system at large with a more or less strong and durable degree of pleasurable sensation, that unlike the, effect of other violently exciting agents, no sensible exhaustion or diminution of vital power accrues from the exertions of its stimulant property, that its most excessive operation even, is neither permanently nor transiently debilitating; and finally, that it fairly promises under judicious application, to prove an extremely efficient remedy, as well in the vast tribe of diseases originating from deficient irritability and sensibility, as in those proceeding from morbid associations, and modifications, of those vital principles.

If you should deem any thing contained in this cursory narrative capable of subserving in any degree the practical advantages likely to result from your scientific and valuable investigation of the genuine properties of the nitrous oxide, it is perfectly at your disposal.

IV. DETAIL OF MR. SOUTHEY

In breathing the nitrous oxide, I could not distinguish between the first feelings it occasioned and an apprehension of which I was unable to divest myself. My first definite sensation was a dizziness, a fulness in the head, such as to induce a fear of falling. This was momentary. When I took the bag from my mouth, I immediately laughed. The laugh was involuntary but highly pleasurable, accompanied by a thrill all through me; and a tingling in my toes and fingers, a sensation perfectly new and delightful. I felt a fulness in my chest afterwards; and during the remainder of the day, imagined that my taste and hearing were more than commonly quick.

---

2. Dr. Mitchill (an American Chemist) has erroneously supported its full admission to the lungs, in its concentrated state, to be incompatible with animal life, and that in a more diluted form it operates as a principal agent in the production of contagious diseases, &c. This gratuitous position is thus unqualifiedly affirmed. "If a full inspiration of gaseous oxyd be made, there will be a sudden extinction of life; and this accordingly accounts for the fact related by Russel (History of Aleppo, p. 232.) and confirmed by other observers, of many persons falling down dead suddenly, when struck with the contagion of the plague." *Vide* Remarks on the Gaseous Oxyd of Azote, by Samuel Latham Mitchill, M.D.

Certain I am that I felt myself more than usually strong and chearful.

In a second trial, by continuing the inhalation longer, I felt a thrill in my teeth; and breathing still longer the third time, became so full of strength as to be compelled to exercise my arms and feet.

Now after an interval of some months, during which my health has been materially impaired, the nitrous oxide produces an effect upon me totally different. Half the quantity affects me, and its operation is more violent; a slight laughter is first induced,[3] and a desire to continue the inhalation, which is counteracted by fear from the rapidity of respiration; indeed my breath becomes so short and quick, that I have no doubt but the quantity which I formerly breathed, would now destroy me. The sensation is not painful, neither is it in the slightest degree pleasurable.

V. LETTER FROM DR. ROGET

The effect of the first inspirations of the nitrous oxide was that of making me vertiginous, and producing a tingling sensation in my hands and feet: as these feelings increased, I seemed to lose the sense of my own weight, and imagined I was sinking into the ground. I then felt a drowsiness gradually steal upon me, and a disinclination to motion; even the actions of inspiring and expiring were not performed without effort: and it also required some attention of mind to keep my nostrils closed with my fingers. I was gradually roused from this torpor by a kind of delirium, which came on so rapidly that the air-bag dropt from my hands. This sensation increased for about a minute after I had ceased to breathe, to a much greater degree than before, and I suddenly lost sight of all the objects around me, they being apparently obscured by clouds, in which were many luminous points, similar to what is often experienced on rising suddenly and stretching out the arms, after sitting long in one position.

---

3. In the former experiments, Mr. Southey generally respired six quarts, now he is unable to consume two. In an experiment made since this paper was drawn up, the effect was rather pleasurable.

I felt myself totally incapable of speaking, and for some time lost all consciousness of where I was, or who was near me. My whole frame felt as if violently agitated: I thought I panted violently: my heart seemed to palpitate, and every artery to throb with violence; I felt a singing in my ears; all the vital motions seemed to be irresistibly hurried on, as if their equilibrium had been destroyed, and every thing was running headlong into confusion. My ideas succeeded one another with extreme rapidity, thoughts rushed like a torrent through my mind, as if their velocity had been suddenly accelerated by the bursting of a barrier which had before retained them in their natural and equable course. This state of extreme hurry, agitation, and tumult, was but transient. Every unnatural sensation gradually subsided; and in about a quarter of an hour after I had ceased to breathe the gas, I was nearly in the same state in which I had been at the commencement of the experiment.

I cannot remember that I experienced the least pleasure from any of these sensations. I can however, easily conceive, that by frequent repetition I might reconcile myself to them, and possibly even receive pleasure from the same sensations which were then unpleasant.

I am sensible that the account I have been able to give of my feelings is very imperfect. For however calculated their violence and novelty were to leave a lasting impression on the memory, these circumstances were for that very reason unfavourable to accuracy of comparison with sensations already familiar.

The nature of the sensations themselves, which bore greater resemblance to a half delirious dream than to any distinct state of mind capable of being accurately remembered, contributes very much to increase the difficulty. And as it is above two months since I made the experiment, many of the minuter circumstances have probably escaped me.

VI. LETTER FROM MR. JAMES THOMSON

The first time I respired nitrous oxide, the experiment was made under a strong impression of fear, and the quantity I breathed not sufficient, as you informed me, to produce the usual effect. I did not note very accurately my sensations. I remember I experienced a slight degree of vertigo after the third or fourth inspiration; and breathed with increased vigor, my inspirations being much deeper and more vehement than ordinary. I was enabled the next time I made the experiment, to attend more accurately to my sensations, and you have the observations I made on them at the time.

After the fourth inspiration, I experienced the same increased action of the lungs, as in the former case. My inspirations became uncommonly full and strong, attended with a thrilling sensation about the chest, highly pleasurable, which increased to such a degree as to induce a fit of involuntary laughter, which I in vain endeavoured to repress. I felt a slight giddiness which lasted for a few moments only. My inspirations now became more vehement and frequent; and I inhaled the air with an avidity strongly indicative of the pleasure I received. That peculiar thrill which I had at first experienced at the chest, now pervaded my whole frame; and during the two or three last inspirations, was attended with a remarkable tingling in my fingers and toes. My feelings at this moment are not to be described; I felt a high, an extraordinary degree of pleasure, different from that produced by wine, being divested of all its gross accompaniments, and yet approaching nearer to it than to any other sensation I am acquainted with.

I am certain that my muscular strength was for a time much increased. My disposition to exert it was such as I could not repress, and the satisfaction I felt in any violent exertion of my legs and arms is hardly to be conceived. These vivid sensations were not of long duration; they diminished insensibly, and in little more than a quarter of an hour I could perceive no difference between the state I was then in, and that previous to the respiration of the air.

The observations I made on repeating the experiment, do not differ from the preceding, except in the circumstance of the involuntary laughter, which I never afterwards experienced, though I breathed the air several times; and in the following curious fact, which, as it was dependent on circumstances, did not always occur.

Having respired the same quantity of air as usual, and with precisely the same effects, I was surprised to find myself affected a few minutes afterwards with the recurrence of a pain in my back and knees, which I had experienced the preceding day from fatigue in walking. I was rather inclined to deem this an accidental coincidence than an effect of the air; but the same thing constantly occurring whenever I breathed the air, shortly after suffering pain either from fatigue, or any other accidental cause, left no doubt on my mind as to the accuracy of the observation.

I have now given you the substance of the notes I made whilst the impressions were strong on my mind. I cannot add any thing from recollection that will at all add to the accuracy of this account, or assist those who have not respired this air, in forming a clearer idea of its extraordinary effects. It is extremely difficult to convey to others by means of words, any idea of particular sensations, of which they have had no experience. It can only be done by making use of such terms as are expressive of sensations that resemble them, and in these our vocabulary is very defective. To be able at all to comprehend the effects of nitrous oxide, it is necessary to respire it, and after that, we must either invent new terms to express these new and particular sensations, or attach new ideas to old ones, before we can communicate intelligibly with each other on the operations of this extraordinary gas.

VII. DETAIL OF MR. COLERIDGE

The first time I inspired the nitrous oxide, I felt an highly pleasurable sensation of warmth over my whole frame, resembling that which I remember once to have experienced after returning from a walk in the snow into a warm room. The only motion which I felt inclined to make, was that of laughing at those who were looking at me. My eyes felt distended, and towards the last, my heart beat as if it were leaping up and down. On removing the mouth-piece the whole sensation went off almost instantly.

The second time, I felt the same pleasurable sensation of warmth, but not I think, in quite so great a degree. I wished to know what effect it would have on my impressions; I fixed my eye on some trees in the distance, but I did not find any other effect except that they became dimmer and dimmer, and looked at last as if I had seen them through tears. My heart beat more violently than the first time. This was after a hearty dinner.

The third time I was more violently acted on than in the two former. Towards the last, I could not avoid, nor indeed felt any wish to avoid, beating the ground with my feet; and after the mouth-piece was removed, I remained for a few seconds motionless, in great extacy.

The fourth time was immediately after breakfast. The few first inspirations affected me so little that I thought Mr. Davy had given me atmospheric air: but soon felt the warmth beginning about my chest, and spreading upward and downward, so that I could feel its progress over my whole frame. My heart did not beat so violently; my sensations were highly pleasurable, not so intense or apparently local, but of more unmingled pleasure than I had ever before experienced.[4]

VIII. DETAIL OF MR. WEDGWOOD

July 23, I called on Mr. Davy at the Medical Institution, who asked me to breathe some of the nitrous oxide, to which I consented, being rather a sceptic as to its effects, never having seen any person affected. I first breathed about six quarts of air which proved to be only common atmospheric air, and which consequently produced no effect.

I then had 6 quarts of the oxide given me in a bag undiluted, and as soon as I had breathed three or four respirations, I felt myself affected and my respiration hurried, which effect increased rapidly until I became as it were entranced, when I threw the bag from me and kept

---

4. The doses in these experiments were from five to seven quarts.

breathing on furiously with an open mouth and holding my nose with my left hand, having no power to take it away though aware of the ridiculousness of my situation. Though apparently deprived of all voluntary motion, I was sensible of all that passed, and heard every thing that was said; but the most singular sensation I had, I feel it impossible accurately to describe. It was as if all the muscles of the body were put into a violent vibratory motion; I had a very strong inclination to make odd antic motions with my hands and feet. When the first strong sensations went off, I felt as if I were lighter than the atmosphere, and as if I was going to mount to the top of the room. I had a metallic taste left in my mouth, which soon went off. Before I breathed the air, I felt a good deal fatigued from a very long ride I had had the day before, but after breathing, I lost all sense of fatigue.

IX. DETAIL OF MR. GEORGE BURNET

I had never heard of the effects of the nitrous oxide, when I breathed six quarts of it. I felt a delicious tremor of nerve, which was rapidly propagated over the whole nervous system. As the action of inhaling proceeds, an irresistible appetite to repeat it is excited. There is now a general swell of sensations, vivid, strong, and inconceivably pleasurable. They still become more vigorous and glowing till they are communicated to the brain, when an ardent flush overspreads the face. At this moment the tube inserted in the air-bag was taken from my mouth, or I must have fainted in extacy.

The operation being over, the strength and turbulence of my sensations subsided. To this succeeded a state of feeling uncommonly serene and tranquil. Every nerve being gently agitated with a lively enjoyment. It was natural to expect that the effect of this experiment, would eventually prove debilitating. So far from this I continued in a state of high excitement the remainder of the day after two o'clock, the time of the experiment, and experienced a flow of spirits not merely chearful, but unusually joyous.

X. DETAIL OF MR. T. POOLE

A disagreeable sensation as if breaking out into a profuse perspiration, tension of the tympanum, cheeks and forehead; almost total loss of muscular power; afterwards increased powers both of body and mind, very vivid sensations and highly pleasurable. Those pleasant feelings were not new, they were felt, but in a less degree, on ascending some high mountains in Glamorganshire.

On taking it the second time, there was a disagreeable feeling about the face. In a few seconds, the feelings became pleasurable; all the faculties absorbed by the fine pleasing feelings of existence without consciousness; an involuntary burst of laughter.

XI. DETAIL OF MR. HAMMICK

Having never heard any thing of the mode of operation of nitrous oxide, I breathed gas in a silk bag for some time, and found no effects, but oppression of respiration. Afterwards Mr. Davy told me that I had been breathing atmospheric air. In a second experiment made without knowing what gas was in the bag, I had not breathed half a minute, when from the extreme pleasure I felt, I unconsciously removed the bag from my mouth; but when Mr. Davy offered to take it from me, I refused to let him have it, and said eagerly, "let me breathe it again, it is highly pleasant! it is the strongest stimulant I ever felt!" I was cold when I began to respire, but had immediately a pleasant glow extending to my toes and fingers. I experienced from the air a pleasant taste which I can only call sweetly astringent; it continued for some time: the sense of exhilaration was lasting. This air Mr. Davy told me was nitrous oxide. In another experiment, when I breathed a small dose of nitrous oxide, the effects were slight, and sometime afterwards I felt an unusual yawning and languor. The last time that I breathed the gas, the feelings were the most pleasurable I ever experienced; my head appeared light, there was a great warmth in the back and a general unusual glow; the taste was distinguishable for some time as in the former experiment. My ideas were more vivid, and followed the natural order of association. I could not refrain from muscular action.

XII. DETAIL OF DR. BLAKE

Dr. Blake inhaled about six quarts of the air, was affected during the process of respiring it with a slight degree of vertigo, which was almost immediately succeeded by a thrilling sensation extending even to the extremities, accompanied by a most happy state of mind and highly pleasurable ideas. He felt a great propensity to laugh, and his behaviour in some measure appeared ludicrous to those around him. Muscular power seemed agreeably increased, the pulse acquired strength and firmness, but its frequency was somewhat diminished. He perceived rather an unpleasant taste in the mouth and about the sauces for some hours afterwards, but in every other respect, his feelings were comfortable during the remainder of the day.

XIII. DETAIL OF MR. WANSEY

I breathed the gas out of a silk bag, believing it to be nitrous oxide, and was much surprised to find that it produced no sensations. After the experiment, Mr. Davy told me it was common air.

I then breathed a mixture of common air and nitrous oxide. I felt a kind of intoxication in the middle of the experiment, and stopping to express this, destroyed any farther effects.

I now breathed pure nitrous oxide; the effect was gradual, and I at first experienced fulness in the head, and afterwards sensations so delightful, that I can compare them to no others, except those which I felt (being a lover of music) about five years since in Westminster Abbey, in some of the grand chorusses in the Messiah, from the united powers of 700 instruments of music. I continued exhilarated throughout the day, slept at night remarkably sound, and experienced when I awoke in the morning, a recurrence of pleasing sensation.

In another experiment, the effect was still greater, the pulse was rendered fuller and quicker, I felt a sense of throbbing in the head with highly pleasurable thrillings all over the frame. The new feelings were at last so powerful as to absorb all perception. I distinguished during and after the experiment, a taste on the tongue, like that produced by the contact of zinc and silver.

XIV. DETAIL OF MR. RICKMAN

On inhaling about six quarts, the first altered feeling was a tingling in the elbows not unlike the effect of a slight electric shock. Soon afterwards, an involuntary and provoking dizziness as in drunkenness. Towards the close of the inhalation, this symptom decreased; though the nose was still involuntary held fast after the air-bag was removed. The dose was probably an undercharge, as no extraordinary sensation was felt more than half a minute after the inhalation.

XV. DETAIL OF MR. LOVELL EDGWORTH

My first sensation was an universal and considerable tremor. I then perceived some giddiness in my head, and a violent dizziness in my sight; those sensations by degrees subsided, and I felt a great propensity to bite through the wooden mouth-piece, or the tube of the bag through which I inspired the air. After I had breathed all the air that was in the bag, I eagerly wished for more. I then felt a strong propensity to laugh, and did burst into a violent fit of laughter, and capered about the room without having the power of restraining myself.

By degrees these feelings subsided, except the tremor which lasted for an hour after I had breathed the air, and I felt a weakness in my knees. The principal feeling through the whole of the time, or what I should call the characteristical part of the effect, was a total difficulty of restraining my feelings, both corporeal and mental, or in other words, not having any command of one's self.

XVI. DETAIL OF MR. G. BEDFORD

I inhaled 6 quarts. Experienced a sensation of fulness in the extremities and in the face, with a desire and power of expansion of the lungs very pleasurable. Feelings similar to intoxication were produced, without being disagreeable. When the bag was taken away, an involuntary though agreeable laughter took place, and the extremities were warm.

In about a quarter of an hour after the above experiment, I inhaled 8 quarts. The warmth and fulness of the face and extremities were sooner produced during the inspiration. The candle and the persons about me, assumed the same appearances as took place during the effect produced by wine, and I could perceive no determinate outline. The desire and power to expand the lungs was increased beyond that in the former experiment, and the whole body and limbs seemed dilated without the sense of tension, it was as if the bulk was increased without any addition to the specific gravity of the body, which was highly pleasant. The provocation to laughter was not so great as in the former experiment, and when the bag was removed, the warmth almost suddenly gave place to a coldness of the extremities, particularly of the hands which were the first to become warm during the inspiration. A slight sensation of fulness not amounting to pain in the head, has continued for some minutes. After the first experiment, a sensation in the wrists and elbows took place, similar to that produced by the electric shock.

XVII. DETAIL OF MISS RYLAND

After having breathed five quarts of gas, I experienced for a short time a quickness and difficulty of breathing, which was succeeded by extreme languor, resembling fainting, without the very unpleasant sensation with which it is usually attended. It entirely deprived me of the power of speaking, but not of recollection, for I heard every thing that was said in the room during the time; and Mr. Davy's remark "that my pulse was very quick and full." When the languor began to subside, it was succeeded by restlessness, accompanied by involuntary muscular motions. I was warmer than usual, and very sleepy for several hours.

XVIII. LETTER FROM MR. M. M. COATES

I will, as you request, endeavour to describe to you the effect produced on me last Sunday se'nnight by the nitrous oxide, and will at the same time tell you what was the previous state of my mind on the subject.

When I sat down to breathe the gas, I believed that it owed much of its effect to the predisposing agency of the imagination, and had no expectation of its sensible influence on myself. Having ignorantly breathed a bag of common air without any effect, my doubts then arose to positive unbelief.

After a few inspirations of the nitrous oxide, I felt a fulness in my head, which increased with each inhalation, until, experiencing symptoms which I thought indicated approaching fainting, I ceased to breathe it, and was then confirmed in my belief of its inability to produce in me any pleasurable sensation.

But after a few seconds, I felt an immoderate flow of spirits, and an irresistible propensity to violent laughter and dancing, which, being fully conscious of the violence of my feelings, and of their irrational exhibition, I made great but ineffectual efforts to restrain; this was my state for several minutes. During the rest of the day, I experienced a degree of hilarity altogether new to me. For six or seven days afterwards, I seemed to feel most exquisitely at every nerve, and was much indisposed to my sedentary pursuits; this acute sensibility has been gradually diminishing; but I still feel somewhat of the effects of this novel agent.

4

"Oh excellent air bag"

Extracts from two letters written by the future poet laureate ROBERT SOUTHEY in 1799. Southey, a close friend of Davy's, was one of the first to inhale the "delectable air". His comparison of it to "the atmosphere of the highest of all possible heavens" would later adorn the tents and advertising bills of 19th-century laughing gas shows, adding a little poetic flair to the often chaotic proceedings.

TO HIS BROTHER, THOMAS SOUTHEY
12 July 1799

Oh Tom! such a Gas has Davy discovered! the Gazeous Oxyd! oh Tom! I have had some. it made me laugh & tingle in every toe & finger tip. Davy has actually invented a new pleasure for which language has no name. oh Tom! I am going for more this evening — it makes one strong & so happy! So gloriously happy! & without any after debility but instead of it increased strength & activity of mind & body — oh excellent air bag. Tom I am sure the air in heaven must be this wonder working gas of delight.

TO HUMPHRY DAVY
3 August 1799

I have seen nothing of Dr. Roget, and can hear nothing of him: you still, I suppose, go on working with your gaseous oxide, which according to my notions of celestial enjoyment, must certainly constitute the atmosphere of the highest of all possible heavens. I wish I was at the Pneumatic Institution, something to gratify my appetite for that delectable air, and something for the sake of seeing you.

5

"As if by the wand of a wizard entranc'd"

The Pneumatic Revellers (1800), a mock-epic penned by the Cornish clergyman and poet RICHARD POLWHELE. First published in the "ultra Tory" *Anti-Jacobin Review*, the piece is a merciless attack on Beddoes and his lofty claims for the "immortal gas".

THE PNEUMATIC REVELLERS
AN ECLOGUE

" —Trifles, light as air,
Are to the *Theorist* confirmations strong."

SOME PRELIMINARY OBSERVATIONS

Among the variety of philosophical improvements, that distinguish the eighteenth century, a few of the discoveries in *Aërology*, have a just claim to our admiration. But the dexterity with which AIRS are made subservient to medical purposes, is, really, a matter of astonishment. Dr. Beddoes has lately applied the *Gas* of Dr. Priestley — the dephlogisticated nitrous Gas — to the uses of medicine: and the success of this experiment, is such as might have been expected from a man who has advanced, in his scientific researches, with an unparalleled velocity; and who, leaving all his contemporaries behind him, has shewn, how far a philosopher may be carried by the force of a flaming imagination.

That Dr. B. hath "contributed to retard the progress of aëro medical science;" is the cry of those only, who feel their incapacity to follow him in his Career, "*per liquidum artbera,*" affect to prefer, rational investigations to fanciful theories — a dull experiment to a splendid hypothesis. From the "Medical Pneumatic Institution" of Dr. Beddoes, will probable flow the most substantial benefits to mankind. In consequence of their intimacy with airs, our modern sages have promised "great things:" but Dr. B. promises greater still. Dr. Darwin thinks, that, from our Pneumatic acquirements or accomplishments, we shall soon be able to regulate the winds of Heaven, and the waves of the sea — to "ride in the whirlwind and direct the storm!" Yet the attempt to subjugate the Hellespont was accounted no less than madness in Xerxes.

And even in an English monarch, (apt as he was to give himself *airs*) the poor effort to check a wave or two, was deemed presumptuous. So great is the superiority of the moderns to the ancients — so striking are the advances of man — so rapid his strides, at this illuminated era, towards the perfection of his nature!

In the mean time, Mr. Godwin maintains, that we may put off death to as late a period, as we please, by means, to be sure, of the *vital air*; though the philosopher does not so express himself. Dr. B_____s, however, combining in his own great and comprehensive mind, the theories of Darwin and of Godwin; and applying his dephlogisticated nitrous Gas to the purposes of both these philosophers, professes his ability to turn us all into amphibious creatures (as some think, a little out of his own element) — to repair the breaches in our constitutions, whether we have suffered from time or intemperance — to subdue disease and pain — to renovate in the aged, every source of pleasure, and even on earth, to render man immortal. "We shall be sadly disappointed (says Dr. B. in the little tract to which I have just referred my readers) if the Gas do not sometimes prove the most delicious of luxuries, as well as the most salutary of remedies. — That natural or forced decay may be repaired, and the faculty of pleasurable sensation renovated, is no longer a mere conjecture, supported by loose analogies — We see the strongest probabilities daily accumulating in favour of the opinion."—The doctor thus describes the effects of this Gas, on several of his friends.

"The REV. R_____T B_____D felt exhilarated, and was compelled to laugh, not by any ludicrous idea, but by an impulse unconnected with thought, lassitude, and languor through the day afterwards.

"MRS. B_____D, the Children's Friend. At first, pleasurable sensations, occasioning involuntary laughter; some momentary faintness, afterwards. We now understand the regulation of the dose so as, perhaps, to be able to remove Mr. B_____s's languor, and to give Mrs. B. the pleasure, without the transitory faintness.

"MR. R_____T S_____Y could not distinguish between the first effects, and an apprehension of which he was unable to divest himself. His first definite sensations were a fullness and dizziness in the head, such as to induce fear of falling. This was succeeded by a laugh which was involuntary but highly pleasurable, accompanied with a peculiar thrilling in the extremities, — a sensation perfectly new and delightful.

He imagined that his taste and smell were more acute, and is certain that he felt unusually strong and cheerful. He has poetically remarked, that he supposes the atmosphere of the highest of all possible Heavens, to be composed of this Gas."

To DR. B⎯⎯⎯s, himself, on trying the effects of the Gas, the first sensations had nothing unpleasant; the succeeding were agreeable beyond conception. He seemed to himself, at the time, to be bathed, all over, with a bucket full of good humour. A constant fine glow, which affects the stomach, led him, one day to take an inconvenient portion of food, and to try the air, afterwards. It very soon removed the sense of distention. Under a certain administration of the Gas, he thinks, sleep might, possibly, be dispensed with. His morning alertness equals that of a healthy boy. Such stores of health and pleasure, has Dr. B. in reserve for his fellow creatures!

And so wild is my wonder, so intense my gratitude, in the contemplation of a philosopher to whom Newton is an ape, and of a philanthropist to whom Howard is a bear, that I can add no more! Expression is lost in sensation!

THE PNEUMATIC REVELLERS

DR. B_____S
G_____E D_____R
REV. R_____T B_____D
MRS. B_____D, "the Children's Friend"
MR. R_____T S_____Y

SCENE
The Medical Pneumatic Chambers

"Into the heaven of heavens, I have presum'd,
An earthly guest, and drawn empyreal air."

DR. B_____S

My friends! from a world, where disorders are rife,
I call you, to taste of the liquor of life;
A fluid, to render us nimble and fresh,
And purge from its drossy pollution the flesh;
To cherish, each purified body, the blood in,
The spirit of beef, and the essence of pudding;
In short, to convey us, ere long, to the portal
Of heaven, and transform us to beings immortal.
    My Comrades, if Priestley discover'd the gas,
He never could bring such a wonder to pass,
As I just have announc'd:— He could never procure,
With all his importance, a gas that was pure.
Indeed, of the Sage though I e'er was a lover, he
Can scarcely be said to have made the *discovery*.

I hold it, my friends, a position unshaken,
That pure vital air was familiar to Bacon:
And, I think, it was known to the poets and sages
Who liv'd in the classic and fabulous ages;
While the tale of old Dis and Persephone shews
The detection of air in a pink or a rose:
Nay, the story of Eve and the Devil may teach,
That Moses found gas in the bloom of a peach.
If so, the discovery of gas, from the maiden
In Sicily ravish'd, we trace up to Eden:
So, inciting fond Eve to a spiritual revel,
The very first chemist in air, was the devil.

    Yet the substance (alas! We have cause to be serious!)
In Eve effervescing, was damn'd deleterious:
And the gas, in my hands, is salubrious, alone:
By Satan, or Priestley prepar'd, 'tis all one.

    Had I been in Eden, perhaps mother Eve
Would have actually soar'd as she seem'd to believe:

    Albeit, as, instead of ascending, she sunk
Top-heavy, and all her race since, have been drunk;
Tho' late, be it mine the mishap to repair,
And exhibit my pure preparations of air.

    But, ere to inhale it your stomachs I urge,
I'll tell you, in brief, the effects of the purge.

    When I tried it, at first, on a learned society,
Their giddiness seem'd to betray inebriety,
Like grave Mandarins, their heads nodding together;
But afterwards each was as light as a feather:
And they, ev'ry one, cried, 'twas a pleasure extatic;
To drink deeper draughts of the mighty pneumatic.
As if by the wand of a wizard entranc'd,
How wildly they shouted, and gambold, and danc'd:
And e'en as phosphoric their bellies and backs shone;
So strong was the force of the muscular action.

G_____E D_____R
(Drinks; and after a short pause, exclaims)

> Flow more briskly, willowy Cam!
>   I have drawn the nitrous gas:
> O! I know not where I am!
>   Sure, I am not what I was.
>
> On thy Marge while erst I lay,
>   Like thy rushes was my rhyme:
> Raptur'd now I break away
>   With emotions more sublime.
>
> So the lark that, warbling high,
>   Thro' the liquid ether flew!
> B_____s! thus, I bathe in sky,
>   Saturate with ambrosial dew.

REV. R_____T B_____D
(Drinks)

> B_____s! thy living beverage whilst I quaff,
> I laugh — ha, ha — yet know not why I laugh.
> Behold, from these intoxicating vapours,
> R_____t, a pickle-herring, cutting capers!
>   I can't — I can't — O, B_____, what an elf!
> Spite of my reverence — can't — contain myself!
> Now I've a strong desire for further quaffing —
> Ha — ha — ha, ha — I cannot drink for laughing.
> Ha, ha! — Yet, somehow, in this merry mood,
> Creeps o'er my body a strange lassitude.
> My frisky spirits are all spent, at once,
> And in the sad residuum leave — a dunce!"

MRS. B_____D, "the Children's Friend"
(Drinks)

>Oh, I feel a fine sensation
>>Stealing o'er my charmed frame!
>Sweeter far that inhalation,
>>Sweeter than the breath of fame.

>Banish'd every carking care is;
>>Sick disgust, and anxious fear:
>This is, sure, the haunt of fairies!
>>Pleasure, pleasure, wantons here.

>Blithe, as when I skipp'd with Lissy,
>>Crown'd with many a pretty flower,
>B___d___s! how I long to kiss y',
>>In my trembling moonlight bower.

>There, between the opening branches,
>>Stars may shed the silent dew;
>But, upon my heels or haunches,
>>Nectar will I taste with you.

>Yet with sudden qualms I languish;
>>Struggles in my breast the sigh:
>With my transport there is anguish —
>>Doctor! Oh, I faint — I die!

MR. R_____T S_____Y

>I am all nerve! — As from the cup of Circe,
>I shrink, suspicious! — I'm a coward! —
>Poh!

[62]

'Tis but an ague-fit that shakes a Cæsar, *(trembles)*
Gods! I *will* drink! *(drinks)*
My head, my head is dizzy!
At my wit's end, I totter — I shall fall!
No — I am rapt beyond myself — I feel
At my extremities delicious thrillings!
My every sense is exquisitely keen!
My taste is so refin'd, I shall henceforth
Disdain all vulgar viands. — So acute
My Smell, I can, for miles around me, catch
The effluvia rolling thro' the shoreless air,
One vast mephitic sea! — These grosser bodies
I cannot brook. — Thou smooth mahogany!
That with surpassing polish seems to shine
A lustrous plane; and, O ye plates of glass
Sciential, ye are rougher than the ruts
Of waggon wheels! I tremble, as I touch you;
E'en from my delicate fingers ends, thro' all
My frame, too sensitive! I spurn, I spurn
This cumbrous clod of earth; and, borne on wings
Of lady-birds, "all spirit," I ascend
Into the immeasurable space, and cleave
The clear ethereal azure; and from star
To star still gliding, to the heaven of heavens
Aspire, and plunging thro' the sapphire blaze,
Ingulph the dephlogisticated floods
Of life, and riot in immortal Gas!

THE DOCTOR HIMSELF
(Drinks)

    Celestials! — This morning, I own, I was sulky,
    And at dinner I ate, till my body grew bulky.

If at dinner, indeed, I indulge in much merriment,
And dispatch a sirloin, 'tis by way of experiment.
   This, therefore, premising, I now have to tell y',
That in temper a dove, and a sparrow in belly,
To the Gas, which in gaining, the members of some ache,
I owe my complacence and lightness of stomach.
I *float* in a manner — so easy and placid —
The mild milk of kindness absorbs every acid!
   Or rather, of passion subsides the hot tumour,
As all over I'm bath'd with a pail of good-humour!
No languid, no crapular feelings have I —
But as gay as the morn — I'm a boy, I'm a boy!
   Such, such is my fluid, the grand panacea:
Though the public may form a degrading idea
Of my science and zeal, of my labour and trouble,
And judge my fine medical airs — but a bubble!
   And if it be said, that a Doctor and Parson,
In concert together to carry the farce on,
Permit all decorum, appearance, and pomp
To be lost in a Bacchanal dance, or a romp;
If, perchance, it be told, that the smiles and the graces
Of ladies, here languish away in grimaces;
My scheme may be spoil'd; and pneumatics be curst,
And B_____s, in truth, like the bubble, may burst.
   Already, 'tis rumour'd, I'm blown up with vanity,
And give myself airs amid chemic inanity;
And (names that detraction is puffing abroad)
I'm, by turns, a chameleon, a moth, and a toad.
   Lest, therefore, my friends, as we scamper and hop,
The report of this meeting go off in a pop;
Lest the business get wind; — I shall print, with your privity,
An account of the gas, as no matter of levity;
And describe its effects, and their curious congruity
Experienc'd by authors of rare ingenuity,

Who never before, I am certain, had cause
(Though long have they liv'd on the breath of applause)
To rejoice in an air from corruption so free,
As the Gas, my good Sirs, just emitted by me.
 I am sorry, indeed, that a friend in the groupe, here,
After exhilaration complain'd of a stupor;
And that *she*, in her lessons for sucklings, so clever,
Resembled so much an exhausted receiver.
 Yet, soon shall this potent Nepenthe, I trust,
My poor fellow-creatures exalt from the dust;
Inspirit the weary, and banish Ennui,
And rouse from his languor the frail debauchee;
Give muscular power to the palsied and grey,
Nor let trouble 'turn an old man into clay.'
 Perhaps, in my hands, it may shortly preclude
The use or of raiment, of sleep, or of food!
Perhaps, with loud plaudits, the people may own
A discovery to shame the Philosopher's Stone;
When, as my *rare* luxury to taste I exhort all,
I shew what a ninny man is — to be mortal!
 What are *ye*, Rosicrusians! indeed, with your riches,
If, throwing away his light 'thin pair of breeches,'
Thy volatile pupil each country can cross over,
Less cumber'd with rags than the shipwreck'd philosopher;
If the slumber so fleeting, my fellows may need here,
Discredit mattrasses, or couches of Eider;
If the food I create for the palate and paunch
Debar the fond wish for a slice of the haunch,
Bring the gluttons on rich calipashes that revel,
And the soup-meagre cottagers, all to a level;
Discovering the grossness of eating, much shame in,
Quick dissipate every alarm from a famine;
And, as I dispense my pure Gas through the nation,
The corn-business render, a mere speculation!

(ALL *drink again; and dance and sing*)

Then hail, happy days! when the high and the low,
    All nourish'd alike from this air-hospitality,
Shall together with Gas-born benevolence glow,
    And prove, that true bliss must arise from equality;

When, Britons and Gauls! ye shall revel and sing,
    Light, lighter than Gossamers twinkle and glance
Here, thridding a maze, and there link'd in a ring,
    And scarcely touch earth, as ye kindle the dance:

When, finer and finer as waxes your nature;
    Each atom terrene shall fly off from your bodies,
Each particle gross, and, all purified matter,
    Ye shall smell of ambrosia or Gas like a goddess:

Till mounting, as if in balloons, to the sky,
    While pleasure with novel sensations shall strike y',
Through the regions of Gas shall ye flutter and fly,
    A Mercury each man, and each woman a Psyche!

6

"Inflated with supreme intensity"

Extract from Canto I of *Terrible Tractoration!! A Poetical Petition Against Galvanising Trumpery, and the Perkinistic Institution* (1803) by CHRISTOPHER CAUSTIC, a pseudonym for the American writer Thomas Green Fessenden. The book-length poem satirised those in the medical profession, Beddoes included, who opposed the use of the "Perkins Patent Tractors", a much maligned medical tool which Fessenden supported.

CANTO I
OURSELF!

ARGUMENT

Great Doctor Caustic is a sage
Whose merit *gilds* this iron age,
And who deserves, as you'll discover,
When you have conn'd this Canto over,
For grand discoveries and inventions,
A dozen peerages and pensions;
But, having met with rubs and breakers
From Perkins' metal mischief makers;
With but three halfpence in his pocket,
In verses blazing like sky rocket,
He first sets forth in this Petition
His *high* deserts but *low* condition.

From garret high, with cobwebs hung,
The poorest wight that ever sung,
Most gentle Sirs, I come before ye,
To tell a lamentable story.

What makes my sorry case the sadder,
I once stood high on Fortune's ladder;
From whence contrive the fickle Jilt did,
That your Petitioner should be tilted.

And soon the unconscionable Flirt
Will tread me fairly in the dirt,
Unless, perchance, these pithy lays
Procure me *pence* as well as *praise*.

Already doom'd to hard quill-driving,
'Gainst spectred poverty still striving,
When e'er I doze, from vigils pale,
Dame Fancy locks me fast in jail.

Necessity, though I am no wit,
Compels me now to turn a poet;
Not *born*, but *made*, by transmutation,
And chemic process, call'd — *starvation!*

Though Poet's trade, of all that I know,
Requires the least of ready rhino;
I find a deficit of cash is
An obstacle to cutting dashes.

For Gods and Goddesses, who traffic
In cantos, odes, and lays seraphic;
Who erst Arcadian whistle blew sharp,
Or now attune Apollo's Jews-harp,

Have sworn they will not loan me, gratis,
Their jingling sing-song apparatus,
Nor teach me how, nor where to chime in
My *tintinabulum* of rhyming.

What then occurs? A lucky hit —
I've found a substitute for wit;
On Homer's pinions mounting high,
I'll drink Pierian puddle dry.

Beddoes (bless the good Doctor) has
Sent me a bag full of his gas,
Which, snuff'd the nose up, makes wit brighter,
And eke a dunce an airy writer.

With which a brother bard, inflated,
Was so stupendously elated,
He tour'd, like Garnerin's balloon,
Nor stopp'd, like half wits, at the moon.

But scarce had breath'd three times before he
Was hous'd in heaven's high upper story,
Where mortals none but poets enter,
Above where Mahomet's ass dar'd venture.

Strange things he saw, and those who know him
Have said that, in his Epic Poem,
To be complete within a year hence,
They'll make a terrible appearance.

And now, to set my verses going,
Like *Joan of Arc* sublimely flowing,
I'll follow Southey's bold example,
And snuff a sconce full, for a sample.

Good Sir, enough! enough already!
No more, for Heav'n's sake! — steady! — steady!
Confound your stuff! — why how you sweat me!
I'd rather swallow all mount Etna!

How swiftly turns this giddy world round,
Like tortur'd top, by truant twirl'd round;
While Nature's capers wild amaze me,
The beldam's crack'd or Caustic crazy!

I'm larger grown from head to tail
Than Mammoth, elephant, or whale! —
Now feel a 'tangible extension'
Of semi-infinite dimension!

Inflated with supreme intensity,
I fill three quarters of immensity!
Should Phoebus come this way, no doubt,
But I could blow his candle out!

This earth's a little dirty planet,
And I'll no longer help to man it,
But off will flutter, in a tangent,
And make a harum scarum range on't!

Stand ye appall'd! quake! quiver! quail!
For lo I stride a comet's tail!
If my deserts you fail to acknowledge,
I'll drive it plump against your college!!

But if your Esculapian band
Approach my highness, cap in hand,
And show vast tokens of humility,
I'll treat your world with due civility.

As Doctor Young foretold, right soon
I'll make your earth another moon,
And Phoebus then, an arrant ass,
May turn his ponies out to grass.

But now, alas! a wicked wag
Has pull'd away the gaseous bag,
From heav'n, where thron'd, like Jove, I sat,
I'm 'fal'n! fal'n! fal'n' down! flat! flat! flat!

7

"Blown by a rudely malicious blast into a world of reptiles"

Extract from *A Dissertation on the Chymical Properties and Exhilarating Effects of Nitrous Oxide Gas and Its Application to Pneumatick Medicine* (1808), the doctoral thesis of WILLIAM BARTON, a medical student at Philadelphia University. After time as a US Naval Surgeon, in which he helped to modernise the Navy hospital system, Barton went on to become professor of botany at the University of Pennsylvania.

I shall now proceed to give an account of the feelings I have myself experienced, upon respiring this gas. It may not be impertinent to the subject to remark, that the sensations I have described, are not the visionary offspring of a creative brain. So far from being conjured up from the effervescence of fancy, they fall far short of the truth, I believe, in the idea they convey of the pleasurable elevation of mind I have experienced. Indeed, the common expressions by which we designate known feelings or sensations, are but feeble and inadequate vehicles to convey a just conception of such as we have felt upon respiring this gas; and we may say of the pleasures of the nitrous oxide, as Dr. Rush has fancifully though emphatically observed of the pleasures of the moral faculty: "It would require a pen, made of a quill, plucked from an angel's wing, to describe half the pleasures arising from this source."[1] As a simple perception cannot be defined, but must be referred to experience, so these feelings are incapable of correct delineation, and can only be understood by being referred to experiment; for they, like all original sensations, admit of no conception from description, but must be felt to be known; no analogous feelings having previously existed to which they can be compared. The only method by which we can arrive at any accuracy in the description of peculiar and novel sensations, is by employing such language as will best convey an idea of those feelings which are most nearly allied to them. It is in this manner I have endeavoured to paint the particular sensations the gas produced upon me, and though the colouring may be deemed too glowing, yet I can confidently assert that the sensations there described, are not the accidental touches of a creative brush, but the chaste colouring of a correct and original outline. It is, however, sometimes necessary to caricature, in order to render a likeness striking; and this may be done without losing sight, of the peculiar turn which constitutes the trait we wish to imprint. I again aver, that the sensations of which I shall presently give an account, are not the workings of "fancy's witchcraft," but the unsophisticated delineations of the most delightful feelings, which will in others, who

---

1. MSS. Lectures on the pleasures of the senses and the mind.

make the trial, caeteris paribus, never fail to supervene upon the proper introduction of this air into the lungs, in sufficiently large doses.

In January 1807, I first inhaled pure nitrous oxide. I breathed six quarts of it from a bladder.[2] The first inspiration, by which I took about a quart of the air into my lungs, produced no unusual effects on them, owing, I suppose, to its union with the air contained therein. The second, by which I inhaled the whole volume of air contained in the breathing bag, was attended with slight giddiness, and a kind of tranquil pleasurable sensation, accompanied with an impatient eagerness to expel the air from my lungs that I might again experience the same feelings by a new inspiration; this eagerness I manifested by a violent expiration, "that seemed," to use the emphatick words of a by-stander, "as if it would have blown the bladder through." During both these inspirations I was perfectly sensible of my situation, and of my object in breathing from the bladder. When I inhaled the gas a third time, it imparted a saccharine taste, like that of fine cider; my vision became suddenly obscured, so that I had not a distinct perception of the nearest objects. I again felt the same pleasant sensation, previously experienced, with the difference of its being less tranquil. This continued till it produced a pleasurable elixity.[3]

I never before experienced, and of which no words can convey a just idea; but, like all original sensations, it must be experienced to be known. I was affected with a *tinnitus aurium*, which I well recollect to

---

2. This and the succeeding experiment, are those related by Dr. Woodhouse, in his edition of Chaptall's Chymistry (Vol. I. p. 182.) I may, not improperly remark, that the rage he notices, must have been occasioned by the frequent attempts he made to remove the bladder from my mouth while I was breathing; though, at the time, I was only conscious of some intruding power exerting itself to deprive me of my pleasure. In both these experiments, I experienced the most extatical delight, with the exception of this little interruption. The sensations were particularized, in the language in which they now appear, immediately upon my return from the laboratory.
3. I use this word upon the authority of Miss Owenson, who, I believe, coined it; and has, I think, employed it with peculiar elegance and expression in her "Wild Irish Girl."

have continued as long as I was sensible of my situation. A glow was diffused throughout my lungs, and at the same time they were affected with a thrilling or titillating sensation, that afterwards extended itself through every part of my frame; but dwelt longest on the extremities. This sensation, as it respects its effects on the lungs, very much resembles the thrilling or actual vibration [4] induced by the loud blowing of a mail stage horn, in the lungs of a passenger in the close carriage. My lungs felt as if they were dilating, and they continued to impart this sensation of enlargement till I supposed they occupied the whole laboratory with their immensity. I now became totally insensible to the impressions of external things, and the rapturous delight which then entranced my faculties, mars my feeble essay towards its description. This indescribable extacy must be what angels feel; and well might the poetick Southey exclaim upon experiencing it, that "the atmosphere of the highest of all possible heavens must be composed of this gas."

---

4. A "vibration in the lungs," produced by sound, since it is entirely a mechanical impulse or affection, may seem to be an improper mode of expression; but, that different parts of the body may be thus mechanically affected, by certain monotonous sounds and musical tones, I am decidedly of opinion. As the effect of the former, I have myself experienced it, by means of the horn alluded to; and as a consequence of the latter, have voluptuously felt and enjoyed it, while listening to the soul touching tones of the melodious Harmonica, whose mellifluous strains breathed indeed

—"the meaning musick of the heart,
To which, responsive, shakes the varied soul;"

and forcibly impressed me with the divine nature of musick, so beautifully suggested by the feeling Collins, where he personifies it, and thus apostrophizes the created being of his imagination:

"O musick sphere-descended maid,
Friend of pleasure, wisdom's aid."

This mechanical affection of the system is mentioned by Dr. Beattie, who, when speaking of the sources of pleasure derived from melody and harmony, says of

From these extatical sensations of joy, I was aroused by Dr. Woodhouse, who now endeavoured to take the breathing bladder from my mouth. This I obstinately and violently resisted, holding the pipe with great force, between my teeth, and directly began to strike him with frequent blows, which were reiterated with energetick strength, as I was afterwards informed, for I was totally unconscious of any thing that happened during this delirious paroxysm, nor did I recollect it when it was over. The resistance I made was prompted, I suppose, by a sensation I well recollect to have experienced, of some intruding power attempting to remove the cause of my pleasurable inebriety. All my muscles seemed to vibrate, and I felt strong enough to root out mountains and demolish worlds; and, like the spirit of Milton, was "vital in every part." At length I suffered the bag to be taken from me; and as soon as it was removed, felt ten times lighter than the surrounding atmosphere, which prompted a strong and almost irresistible disposition to mount in the air, which I discovered to the spectators by repeatedly jumping up from the floor with great and uncommon agility. My sensations were just such as I should imagine would be produced by flying. I experienced an unrestrainable inclination to muscular motion, opposing much and powerful resistance to all who endeavoured to restrain me. I resembled those varlets, who, as Ariel tells Prospero, in the Tempest,

---

certain inarticulate sounds: "It is not altogether absurd to suppose, that the human body may be mechanically affected by them. If in a church one feels the floor, and the pew, tremble to certain tones of the organ; if one string vibrates of its own accord when another is sounded near it of equal length, tension, and thickness; if a person who sneezes, or speaks loud, in the neighbourhood of a harpsichord, often hears the strings of the instrument murmur in the same tone; we need not wonder, that some of the finer fibres of the human frame should be put in a tremulous motion, when they happen to be in unison with any notes proceeding from external objects." "Essays"— Essay on poetry and musick as they affect the mind, chap, vi. sec. 2. This opinion that a tremulous or vibrating motion in different parts of the human body, may be induced by particular sounds, is corroborated by Dr. Rush, who, in his lecture on voice and speech, says, "the tremours produced in singing are so great, that they are sometimes felt in every part of the body, and some persons have said, that they have felt them in their bones." MSS. *Lectures*

—"were red hot with drinking;
So full of valour, that they smote the air
For breathing in their faces; beat the ground
For kissing of their feet,"[5]

and feeling like the presiding genius of all I beheld, beat with indignant resentment every person that attempted, vainly, as I supposed, to impede my progress. This superiority that I fancied I possessed over all around me, was so ably seconded by my increased muscular strength, that some of the gentlemen who received my blows, told me they were applied with wonderful and disagreeable force. I seemed to be placed on an immense height, and the noise occasioned by the reiterated shouts of laughter and hallooing of the by-standers appeared to be far below me, and resembled the hum or buz which aeronauts describe as issuing from a large city, when they have ascended to a considerable height above it. I had a sense of great fulness and distension in my head, and my thoughts and perceptions, as well as I can recollect, were rapid and confused, but very unlike any I had ever before experienced. By a sensation as sudden as

— "with quick impulse through all nature's frame,
Shoots the electrick air, its subtle flame,"[6]

I seemed to descend from the immense height to which I had flown, and by a quick, but complete prostration of muscular energy, fell into a kind of trance-like state. During the short continuance of this trance, my feelings were placidly delicious, and extremely analogous to those I have often experienced in that state of voluptuous delight, vibrating between a waking consciousness and the torpor of sleep, so elegantly, so feelingly delineated by Rousseau in these words,

---

5. Act 4, scene 1.
6. Darwin's Botanick Garden, Canto 4, line 425.

"Thus lifeless yet with life, how sweet to lie!
Thus without dying oh how sweet to die!"

To this state syncope succeeded, and I was carried into an adjoining room and placed on a table near an open window. Here I experienced a slight return of the agreeable feelings I have before described, but only of instantaneous duration. The first idea that occurred to me upon my partial revival, was a confused one of nitrous oxide, which words I vociferated as I jumped from the table with great vehemence, as I was afterwards informed. I felt much indignation and pride towards the persons around me, and entertained a momentary contempt for every thing that excited an idea in my still chaotick brain. I felt as if I was an inhabitant of the Elysium of Rousseau,[7] or the island of Calypso, of Fenelon,[8] blown by a rudely malicious blast into a world of reptiles, where the atmosphere like the pestiferous samiel[9] of the desarts of Arabia, was pregnant with destruction, and threatened inevitable annihilation to all who inhaled its morbid breath. I now, however, as quick as thought, completely revived, and made the mortifying discovery, that the aerial world through which I had been roving with footsteps light as air, was but the fascination of an inebrieting elixiry, whose siren spell of pleasure wrapt me in delight.

A profuse diaphoresis appeared all over me, but was particularly abundant on my forehead and cheeks; and the temporal arteries both during the experiment and after it was over, seemed ready to burst with fulness.

The next time I breathed the gas, my feelings were, as well as I can recollect, nearly similar to those just described. In this experiment however I experienced one sensation, that I did not feel in the first, *viz.* a kind of titillation in my eyes as if water had been dropping between the ball of the eye and its palpebrae.

---

7. *Vide* St. Preux's beautiful description of this enchanting little spot, in his letter to Lord B_____, Eloisa, letter cxxx. vol. 2.
8. *Vide* Telemachus.
9. *Vide* Lind. on Hot Climates, part 2. chap. 1. sec. 1.

I must not omit to mention here, that I also experienced in this experiment, and in every other that I made except the one just detailed, a sensation extremely singular. It consisted in a kind of semi-consciousness of my situation, yet unattended by perfect volition. Thus I became enraged as in the preceding experiment, at the vain presumption, as I deemed it, of those who dared to oppose my motions, supposing them my antagonists; at the same time I seemed to be sensible they were not so, and conceived myself under the influence of some incomprehensible hallucination, the effect of which, however, I was unable to resist, and of consequence combated with them against my will. I seemed as it were, to have two kinds of consciousness, the one persuading me that I was actually opposed by enemies, the other rendering me sensible, that this was entirely a misconception of the obvious reality, which was that my enemies were indeed no other than friendly spectators, and that their actions which were ostensibly inoffensive, I had misconstrued into the exertion of violence and power against me. Volition, however, was wholly inactive, or, if I may be allowed the expression, paralized; of consequence I derived no benefit from the effect of its operations. I may perhaps illustrate this semi-conscious, semi-delusive state, of which notwithstanding my efforts to describe it, I feel unable to convey a just conception, by the following description of an analogous situation, by the celebrated Kotzebue. It occurred to him during the night after his arrival at Tobolsk, after a fatiguing, an anxious, and distressing journey; he had been, perhaps, affected with a disordered state of his mind, induced by the contemplation of a melancholy exile in the chills of Siberia, separated from his beloved family. "In the course of the night," says he, "a remarkable circumstance took place, the explanation of which I must leave to my good friends, doctors Gall and Hufeland. I had fallen asleep; towards twelve o'clock I awoke, and fancied myself on board a ship. I not only felt the rocking motion of the vessel, but heard the flapping of the sails, and the noise and bustle of the crew. As I lay on the floor, I could see no objects through the window, except the sky, and this circumstance added to the force of the illusion. I was sensible it was such, and endeavoured to overcome it. I felt myself, as it were, furnished with two separate minds, the one confirmed what I fancied, the other convinced me that it was all imaginary. I staggered about the room, thought I saw the counsellor,[10] and every thing that surrounded me the evening before, remaining in the same place. I went to the window; the wooden houses in the streets I thought were ships, and in every direction I perceived the open sea. Whither am I going? seemed to say one mind. Nowhere, replied the other; you are still in your own apartment. This singular sensation, which I cannot well describe, continued for half an hour; by degrees it became less powerful, and at length entirely quitted me. A violent palpitation of the heart, and a quick convulsive pulse succeeded. Yet I was not feverish, nor did I feel any headach. My own opinion and conviction is, that the whole must have been the commencement of a species of insanity."[11]

DETAIL OF SOME EXPERIMENTS PERFORMED BEFORE
THE CHYMICAL CLASS, IN DECEMBER, 1807

With a view to satisfy some few gentlemen who were still skeptical as to the reality of the effects of nitrous oxide, but more particularly that I might ascertain its effects upon the pulse, I determined again to inhale it once or twice in the presence of the class. I also attended to the state of the pulse, both previous and subsequent to the experiments made by some other gentlemen, of which I shall presently give an account.

DEC. 7. At eleven o'clock, my pulse being at its natural standard 96, and my mind undisturbed, I inhaled five quarts of the gas, having previously made as complete an exhaustion of the air of my lungs as I could effect. The gas was sudden in its operation, and I recognised its sweetish taste as soon as it came in contact with the fauces. My pleasure was less sublime, but more lively than in any former experiments. I had an

---

10. The Aulick Counsellor, who had been his escort and guard into Siberia, and who had then left him.
11. Kotzebue's Life, written by himself, vol. 1. p. 256.

intense and vivid recollection of the delightful sensations I had before experienced while breathing the gas[12] — felt a strong inclination to express my delight by speech, and recollect to have found my language incapable of conveying an idea of my pleasure, though I was told that I repeatedly exclaimed with rapture and extacy, "Oh if such is heaven, then indeed it is desirable." I experienced more of the tingling in the extremities than on former occasions, and the sensation was extremely pleasurable. My feet tottered under me, as I well recollect, and I fainted, but soon recovered upon being removed into an adjoining room near an open window. Upon my revival I became indignant as before, and beat every one who approached me.

---

12. This is a proof of Dr. Rush's position, that the recalling of ideas is owing to the same motion which originally produced a particular idea, being again reproduced in the same place. MSS. Lectures. — Lect. on the mind.

8

"Dancing, jumping, kicking, fencing, and occasionally boxing"

Extract from *A Cursory Glimpse of the State of the Nation, on the Twenty-Second of February, 1814 ... or, A Physico-Politico-Theologico Lucubration Upon the Wonderful Properties of Nitrous Oxide* (1814) by the Philadelphia-based printer and pamphleteer MOSES THOMAS. After a lively description, reproduced here, of a laughing gas "lecture", the bizarre treatise segues into a lengthy and colourful rant upon the dangers of the US invading Canada.

Nitrous oxide supports combustion — a taper placed in it burns with considerable brilliancy — other combustibles are similarly affected, but its most distinguishing property is, its effect upon the human system, when inhaled by the mouth — an effect so singular, and so powerful, that it is still witnessed with astonishment, even by those who have had the most frequent opportunities of observing it; the exhilarating article being applied to an organ (the lungs) through which no such effect could be apprehended. It is at the same time so delightful, and passes off so suddenly, that it seems more like the effects ascribed to enchantment, than those producible by the intervention of any natural agent.

Passing a leisure moment, the other evening, at the WASHINGTON hotel in Sixth-street, for the taverns and coffee-houses of the days of Addison and Steele, are with us converted into inns and hotels, and happening to cast my eye over Relf's Philadelphia Gazette, I chanced to observe that Dr. Jones's weekly lecture upon this interesting subject, was advertised for the last time this season. I immediately called for my hat and cane, and sallied forth to procure a ticket, and to inquire for Harmony court, at the corner of which, it seems, the learned doctor exhibits his supernatural experiments.

The lecture room is an oblong of twenty feet by thirty, one end of which is separated from the physical apparatus, by a transverse writing-desk, behind which rise a dozen benches, in regular gradation, the entrance to which is barred across, to prevent the inhalers of the gas from too ready access to the ladies; who are advised, as they enter, to place themselves upon the hindmost seats — that they may be out of harms way. When the doctor has descanted, at sufficient length, upon the nature and properties of the nitrous oxide; and exhibited a number of unimportant experiments, to which very little attention is paid by his auditors, who come rather to see — than to hear; he begins to perceive the impatience, particularly of the female part of the company, and he proposes to deliver ten or twelve tickets, regularly numbered, to so many young gentlemen who may have a mind to inhale the exhilarating gas. The pit is now cleared for action, and the first on the list, stepping eagerly forward (if he has ever taken it before) receives a large bladder, inflated with the proper portion of nitrous oxide.

On the present occasion the first practitioner was a fine youth of fifteen, who inhaled the gas with spirited avidity — suddenly threw away the bag, with an air of triumphant disdain, and began to march about the inclosure with theatric strides, until coming close up to the front row, he perceived that one of the persons who sat there held a cane athwart to defend himself from his too near approach. This offended his pride — he instantly burst into a paroxysm of rage: "That tyrant!" says he, "has seized my cane — deliver it to me! — this — instant! — or — I'll be the death of you!" At the same moment jumping over the desk, and grappling with the man who had the cane, he overturned every thing that stood in his way, and it required the united efforts of four or five men to hold him down, till the effect of the gas ceased, and he turned round to the company with an air of good-humoured hilarity.

Several others now trod the stage, in turn, with different degrees of animation, or ferocity, dancing, jumping, kicking, fencing, and occasionally boxing any one that stood in their way; when a young man of five and twenty approached the table, inhaled a potent dose of the delicious poison, and began to display its effects upon his frame, by dashing at the candles — driving off the doctor — and, finally, advancing to the company, he threw himself into the most haughty attitude he could assume, and exclaimed, with terrifying emphasis, "Byy hea-vens! — 'Twere nobly done! — To snatch the briidal honours — from the blaazing sunn!" This violent exertion exhausted the draught he had inhaled. He turned about as if amazed, and sat quietly down upon a bench that was near him.

I do not recollect any thing more observable, in those that followed, than that an ingenious boy, after amusing the company by his freakish activity, turned suddenly to the doctor, and offered him his hand, saying, "Well, doctor, here I am, at last;" as if he had just come off of a journey, and was glad to see his friends again. Though one sprightly youth danced rapidly round the ring, aiming a kick at one — giving a slap on the face to another — and shaking his fist, at a third; till, finally, throwing himself headlong into the midst of his supposed enemies, he struggled with them for a moment; and then instantly came to himself, without having spoken a single word throughout the whole pantomime: for it is observable on this confined theatre, as well as in that of real life, that the greatest fighters are — men of few words — and no pretensions.

A powerful young man of six foot, now offered himself at the table, upon which most of those who were on the bench below me decamping, I also thought it most prudent to get out of the way of the first onset, as there was no knowing how furious it might be. He had by this time inhaled his potion, with the most evident signs of delight, and was marching, or rather stamping, along the boards, when he suddenly assumed a fixed posture — faced the company — and with uplifted hands and eyes exclaimed — in a voice of thunder — "Alexaander!!!" — This exhausted his strength, and as he fell to the floor, like a log, he cried out, "Lord! deliver us!"

The exhilarating gas was now spent, and I could not but then compare the theatric rhapsodies to which I had been a witness, to the bombastic effusions of our western generals on entering Canada — since the parallel held out so exactly in their falling away, as the supernatural vigour excited by the inflating gas, exhausted itself in fumo, leaving upon the escutcheon of their offended country, that stain of pusillanimity, which the blood of LAWRENCE, and ALLEN, and BURROWS, and so many more of our gallant seamen (the last relics of THE WASHINGTON POLICY) afterward flowed to wash away: for they have been truly said to be obliged to fight their way to favour with the present professedly economical Administration.

But on my return to my chambers, and when I laid myself down to sleep, between sleeping and waking, I carried the comparison further. It appeared to me as though the United States of America, in congress assembled, had inhaled an imprudent portion of the exhilarating gas, which they were now actually breathing forth again — in defiance of God and man. The gallant youth, who swore that the tyrant had got his cane, and that he would be the death of him, but he would have it again, reminded me of our pertinacious determination to have every thing yielded up to us that we contend for. The blustering bravado of the young man that cried,

"'Twere nobly done! —
To snatch the bridal honours from the blazing sun!'"

appeared to tally with sufficient exactness to our occasional threats to sweep every sea, and exclude the navy of Great Britain from the ocean. And the pathetic exclamation of him who invoked the name of ALEXANDER, as he was falling to the ground, with utter imbecility, bore too striking an allusion to be mistaken to our flattering prospect of an eventual accommodation, through the friendly interference of the Deliverer of Europe. —

It grieves me to expose the nakedness of MY COUNTRY — in a state of political intoxication; and I would not — unnecessarily — hurt the feelings of the least of her well meaning public functionaries. — If I have probed to the quick, the wounds of the daughter of my people, and laid open her bruises, and her putrifying sores; it is not to aggravate — but to heal: "Faithful are the wounds of a friend." — [Proverbs xxvii. 6.]

These apparently "dazzling miracles," however to recur to the motto of my paper, are yet capable of an easy solution, for nothing can be more natural than that the minds of young men, in a state of inconceivable excitation, should turn upon the recollected injuries of their country — its adventurous expeditions — or the inspiring prospects of Its returning prosperity. Yet the probability of this singular chain of historical coincidences may well be doubted by others: for I can now scarcely credit my own recollection of it. The whole story, I well know, will be supposed to be nothing more than a waking dream, a political vision: but I have only to refer the sceptic to any one of more than a hundred persons, of both sexes, who were present at the exhibition I have described, for proof that the facts occurred, and that too, in the very order in which I have related them, on the identical evening of the 23rd of February.

9

"This is a queer world"

The finale to *The Anaesthetic Revelation and the Gist of Philosophy* (1874), a privately published 37-page pamphlet produced by BENJAMIN PAUL BLOOD, a philosopher and poet from New York. While the author of several books, and known during his time for his poetry, Blood's major output was epistolary, in the form of letters to local newspapers and correspondence with friends including Alfred Lord Tennyson and fellow nitrous inhaler and philosopher William James.

By the Anaesthetic Revelation I mean a certain survived condition, (or uncondition,) in which is the satisfaction of philosophy by an appreciation of the genius of being, which appreciation cannot be brought out of that condition into the normal sanity of sense — cannot be formally remembered, but remains informal, forgotten until we return to it.

"As here we find in trances, men
Forget the dream, that happens then,
Until they fall in trance again."

Of this condition, although it may have been attained otherwise, I know only by the use of anaesthetic agents. After experiments ranging over nearly fourteen years I affirm — what any man may prove at will — that there is an invariable and reliable condition (or uncondition) ensuing about the instant of recall from anaesthetic stupor to sensible observation, or "coming to," in which the genius of being is revealed; but because it cannot be remembered in the normal condition it is lost altogether through the infrequency of anaesthetic treatment in any individual's case ordinarily, and buried, amid the hum of returning common sense, under that epitaph of all illumination: "this is a queer world." Yet I have warned others to expect this wonder on entering the anaesthetic slumber, and none so cautioned has failed to report of it in terms which assured me of its realization. I have spoken with various persons also who induce anesthesis professionally (dentists, surgeons, etc.,) who had observed that many patients at the moment of recall seem as having made a startling yet somehow matter-of-course (and even grotesque) discovery in their own nature, and try to speak of it, but invariably fail in a lost mood of introspection. Of what astonishes them it is hard to give or receive intimation; but I think most persons who shall have tested it will accept this as the central point of the illumination: That sanity is not the basic quality of intelligence, but is a mere condition which is variable, and like the humming of a wheel, goes up or down the musical gamut according to a physical activity; and that only in sanity is formal or contrasting thought, while the naked life is realized only outside of sanity altogether;

and it is the instant contrast of this "tasteless water of souls" with formal thought as we "come to" that leaves in the patient an astonishment that the awful mystery of Life is at last but a homely and a common thing, and that aside from mere formality the majestic and the absurd are of equal dignity. The astonishment is aggravated as at a thing of course, missed by sanity in overstepping, as in too foreign a search, or with too eager an attention: as in finding one's spectacles on one's nose, or in making in the dark a step higher than the stair. My first experiences of this revelation had many varieties of emotion; but as a man grows calm and determined by experience in general, so am I now not only firm and familiar in this once weird condition, but triumphant — divine. To minds of sanguine imagination there will be a sadness in the tenor of the mystery, as if the key-note of the universe were low, — for no poetry, no emotion known to the normal sanity of man can furnish a hint of its primeval prestige, and its all-but appalling solemnity; but for such as have felt sadly the instability of temporal things there is a comfort of serenity and ancient peace; while for the resolved and imperious spirit there are majesty and supremacy unspeakable. Nor can it be long until all who enter the anaesthetic condition (and there are hundreds every secular day) will be taught to expect this revelation, and will date from its experience their initiation into the Secret of Life.

Men and brethren, into this pervading genius we pass, forgetting and forgotten, and thenceforth each is all, in God. There is no higher, no deeper, no other, than the life in which we are founded.

"The One remains, the many change and pass;"

and each and every of us is the One that remains. — Listen, then, to the charming of the Prince of Peace, who takes away the sin of the world, and say, each for himself, "My Father and I are one." — Mourn not for the dead, who have awoke in the bosom of God. They care not, they think not, and when we are what they are, we too shall think of them no more. — Much might I say of the good of this discovery, if it were, as it soon may be, generally known of. Now for the first time the ancient problem is referred to empirical resolution, when the expert and the novice may meet equally on the same ground. My worldly tribulation reclines on its divine composure; and though not in haste to die, I "care not to be dead," but look into the future with serene and changeless cheer. This world is no more that alien terror which was taught me. Spurning the cloud-grimed and still sultry battlements whence so lately Jehovan thunders boomed, my gray gull lifts her wing against the nightfall, and takes the dim leagues with a fearless eye.

By this revelation we enter to the sadness and the majesty of Jesus — to the solemn mystery which inspired the prophets of every generation. By some accident of being they entered to this condition. This is "the voice of One crying in the wilderness, Make straight the way of the Lord." He that hath ears to hear let him hear. Heed not for themselves the voice nor the hand, which ever deny themselves; remember only how many inspired times it is spoken and written: I AM — that God whom faltering spirits seek in far-off courts of heaven, while behold! the kingdom of God is neither "lo! here" nor "lo! there" but within you; it is the Soul. Thou shalt vanish, but the Soul is eternal: I speak not of souls. And behold, I say unto you, the Supreme Genius doth not facultize; the glory is not what it does but what it is; it hath no old nor new, no here nor there; it stays not to remember, to wonder, to compare; to the vehm of the patrician Presence, omniscience were an idle labor and delay, and prophecy is forestalled and bootless in the sole sufficiency whose paean hath no echo.

This is the Ultimatum. It is no glance between conditions, as if in passing from this sphere of existence we might catch a glimpse of

> "The Gods, who haunt
> The lucid interspace of world and world,
> Where never creeps a cloud, or moves a wind,
> Nor ever falls the least white star of snow,"

and lose them again as we pass on to another orb and organization. This thick net of space containing all worlds — this fate of being which contains both gods and men, is the capacity of the Soul, and can be claimed as greater than us only by claiming a greater than the greatest, and denying God and safety. As sure as being — whence is all our care — so sure is content, beyond duplexity, antithesis, or trouble, where I have triumphed in a solitude that God is not above.

It is written that "there was war in heaven," — that aeons of dominion, as absolute as any, beheld the banners of Lucifer streaking with silver and crimson the mists of the morning, and heard the heavy guns of Moloch and Belial beating on the heights of the mind; and I read that dead men have appeared as human forms; — nought of this can I deny, more or better than I can deny myself. The tales, whether they be true or false, are as substantial as the things of which they tell.

> "We are such stuff
> As dreams are made of, and our little life
> Is rounded with a sleep."

10

"Good and evil reconciled in a laugh!"

The afternote to "On Some Hegelisms", an essay by the philosopher WILLIAM JAMES, which first appeared in the journal *MIND*, Vol. VII, 1882. Twenty years later, James would publish his seminal work *The Varieties of Religious Experience* (1902), the central idea of which — the psychological truth of "extraordinary consciousness" — he traced back directly to his experiences with nitrous oxide.

…I have made some observations on the effects of nitrous-oxide-gas-intoxication which have made me understand better than ever before both the strength and the weakness of Hegel's philosophy. I strongly urge others to repeat the experiment, which with pure gas is short and harmless enough. The effects will of course vary with different individuals just as they vary in the same individual from time to time; but it is probable that in the former case as in the latter a generic resemblance will obtain. With me, as with every other individual of whom I have heard, the keynote of the experience is the tremendously exciting sense of an intense metaphysical illumination. Truth lies open to the view in depth beneath depth of almost blinding evidence. The mind sees all the logical relations of being with an apparent subtlety and instantaneity to which its normal consciousness offers no parallel; only as sobriety returns, the feeling of insight fades, and one is left staring vacantly at a few disjointed words and phrases, as one stares at a cadaverous-looking snow peak from which the sunset glow has just fled, or at the black cinder left by an extinguished brand.

The immense emotional sense of *reconciliation* which characterizes the "maudlin" stage of alcoholic drunkenness,—a stage which seems silly to lookers-on, but the subjective rapture of which probably constitutes a chief part of the temptation to the vice, is well known. The centre and the periphery of things seem to come together. The ego and its objects, the *meum* and the *tuum*, are one. Now this, only a thousandfold enhanced, was the effect upon me of the gas: and its first result was to make peal through me with unutterable power the conviction that Hegelism was true after all, and that the deepest convictions of my intellect hitherto were wrong. Whatever idea or representation occurred to the mind was seized by the same logical forceps, and served to illustrate the same truth, and that truth was that every opposition, among whatsoever things, vanishes in a higher unity in which it is based; that all contradictions, so called, are but differences; that all differences are of degree; that all degrees are of a common kind; that unbroken continuity is of the essence of being; and that we are literally in the midst of *an infinite*, to perceive the existence of which is the utmost we can attain. Without the *same* as a basis, how could strife occur? Strife presupposes something to be striven about; and in this common topic, the same for both parties, the differences merge. From the hardest contradiction to the tenderest diversity of verbiage differences evaporate, *yes* and *no* agree at least in being assertions, — a denial of a statement is but another mode of stating the same, contradiction can only occur of the same thing, all opinions are thus synonyms, are synonymous, are the same. But the same phrase by difference of emphasis is two; and here again difference and no-difference merge in one. It is impossible to convey an idea of the torrential character of the identification of opposites as it streams through the mind in this experience. I have sheet after sheet of phrases dictated or written during the intoxication, which to the sober reader seem meaningless drivel, but which at the moment of transcribing were fused in the fire of infinite rationality. God and devil, good and evil, life and death, I and thou, sober and drunk, matter and form, black and white, quantity and quality, shiver of ecstasy and shudder of horror, vomiting and swallowing, inspiration and expiration, fate and reason, great and small, extent and intent, joke and earnest, tragic and comic, and fifty other contrasts figure in these pages in the same monotonous way. The mind saw how each term *belonged* to its contrast through a knife-edge moment of transition which *it* effected, and which, perennial and eternal, was the *nunc stans* of life. The thought of mutual implication of the parts in the bare form of a judgment of opposition, as "nothing — but," "no more — than," "only — if," &c., &c., produced a perfect delirium of theoretic rapture. And at last, when definite ideas to work on came slowly, the mind went through the mere form of recognizing sameness in identity by contrasting the same word with itself, differently emphasised, or shorn of its initial letter. Let me transcribe a few sentences:

> What's mistake but a kind of take?
> What's nausea but a kind of -ausea?
> Sober, drunk, -*unk*, astonishment.
> Everything can become the subject of criticism —
>   how criticise without something to criticise?
> Agreement — disagreement!!

Emotion — motion!!!

Die away from, *from*, die away (without the *from*).

Reconciliation of opposites; sober, drunk, all the same!

Good and evil reconciled in a laugh!

It escapes, it escapes!

But ——

What escapes, WHAT escapes?

Emphasis, EMphasis; there must be some emphasis in order for there to be a phasis.

No verbiage can give it, because the verbiage is *other*.

*In*coherent, coherent — same.

And it fades! And it's infinite! AND it's infinite!

If it wasn't *going*, why should you hold on to it?

Don't you see the difference, don't you see the identity?

Constantly opposites united!

The same me telling you to write and not to write!

Extreme — extreme, extreme! Within the extensity that "extreme" contains is contained the "*extreme*" of *in*tensity.

Something, and *other* than that thing!

Intoxication, and *otherness* than intoxication.

Every attempt at betterment, — every attempt at otherment, — is a ——.

It fades forever and forever as we move.

There *is* a reconciliation!

Reconciliation — econciliation!

By God, how that hurts! By God, how it *doesn't* hurt! Reconciliation of two extremes.

By George, nothing but othing!

That sounds like nonsense, but it is pure *on*sense!

Thought deeper than speech —— !

Medical school; divinity school, *school!* SCHOOL! Oh my God, oh God, oh God!

There are no differences but differences of degree between different degrees of difference and no difference. &c., &c., &c.

This phrase has the true hegelian ring, being in fact a regular *sich als sich auf sich selbst beziehende Negativität*. And true Hegelians will *überhaupt* be able to read between the lines and feel, at any rate, what possible ecstasies of cognitive emotion might have bathed these tattered fragments of thought when they were alive. But for the assurance of a certain amount of respect for them, I should hardly have ventured to print what must be such caviare to the general.

But now comes the reverse of the medal. What is the principle of unity in all this monotonous rain of instances? Although I did not see it at first, I soon found that it was in each case nothing but the abstract *genus* of which the conflicting terms were opposite species. In other words, although the flood of ontologic *emotion* was hegelian through and through, the *ground* for it was nothing but the world-old principle that things are the same only so far and no farther than they are the same, or partake of a common nature — the principle that Hegel most tramples under foot. At the same time the rapture of beholding a process that was infinite changed, as the nature of the infinitude was realised by the mind, into the sense of a dreadful and ineluctable fate, with whose magnitude every finite effort is incommensurable and in the light of which whatever happens is indifferent. This instantaneous revulsion of mood from rapture to horror is, perhaps, the strongest emotion I have ever experienced. I got it repeatedly when the inhalation was continued long enough to produce incipient nausea, and I cannot but regard it as the normal and inevitable outcome of the intoxication, if sufficiently prolonged. A pessimistic fatalism, depth within depth of impotence and indifference, reason and silliness united, not in a higher synthesis, but in the fact that whichever you choose it's all one, — this is the upshot of a revelation that began so rosy bright.

Even when the process stops short of this ultimatum the reader will have noticed from the phrases quoted how often it ends by losing the clue. Something "fades," "escapes," and the feeling of insight is changed into an intense one of bewilderment, puzzle, confusion, astonishment. I know no more singular sensation than this intense bewilderment, with nothing particular left to be bewildered at, save the bewilderment itself. This seems the true *causa sui*, or "spirit become its own object".

My conclusion is that the togetherness of things in a common world, the law of sharing, of which I have said so much, may, when perceived, engender a very powerful emotion; that Hegel was so unusually susceptible to this emotion throughout his life that its gratification became his supreme end and made him tolerably unscrupulous as to the means he employed; that *indifferentism* is the true outcome of every view of the world which makes infinity and continuity to be its essence, and that pessimistic or optimistic attitudes pertain to the mere accidental subjectivity of the moment; finally that the identification of contradictories, so far from being the self-developing process which Hegel supposes, is really a self-consuming process, passing from the less to the more abstract, and terminating either in a laugh at the ultimate nothingness or in a mood of vertiginous amazement at a meaningless infinity.

11

"Om! Om! Om! Om! Om! Om! Om!"

"Laughing Gas", a short one-act play by the American writer THEODORE DREISER, part of his *Plays of the Natural and Supernatural* (1916). Best known for his novels, and a nominee for the Nobel Prize in Literature in 1930, Dreiser once referred to the play as "the best thing I ever did". It was written in the summer of 1914, inspired by Dreiser's own surgical experience earlier in the year.

LAUGHING GAS

CHARACTERS:

JASON JAMES VATABEEL	An eminent physician
FENWAY BAIL	A celebrated surgeon
ARTHUR GAILEY	House physician of the Michael Slade Memorial Hospital
SLASON TUFTS	His assistant
FRANKLIN DRYDEN	An anesthetist
DEMYAPHON	Nitrous oxide, an element of chemistry
ALCEPHORAN	A power of physics

SHADOWS AND VOICES of the first, second, third and fourth planes
NURSES AND INTERNES of the Michael Slade Memorial Hospital
THE RHYTHM OF THE UNIVERSE

SCENE:

The operating-room of the Michael Slade Hospital, a glistening chamber of white porcelain and white tile. Nickel operating table in the foreground. Racks of surgical implements and supplies to either side. A strong, even light from the north French windows. Attendants in white bustling about preparatory to an operation. Enter FENWAY BAIL, *an eminent surgeon, and* JASON JAMES VATABEEL, *his friend, a celebrated physician. They are followed by* ARTHUR GAILEY, *chief house physician;* SLASON TUFTS, *his assistant;* FRANKLIN DRYDEN, *the anesthetist, and two nurses.*

BAIL *(A cool, sallow-faced, collected man of perhaps fifty-five, wise and incisive.)* Well, Jason, here you are, a victim of surgery after all!

VATABEEL *(Tall, gaunt, all of fifty-eight, very distinguished, a little pale from recent suffering, a bandage about his neck, beginning to loosen his shirt in front.)* The last time I took ether I had a very strange experience or dream, one of the best of the etheric variety, I fancy. I am wondering whether it will repeat itself today.

BAIL *(Examining a case of instruments, and busy with asides to Gailey and others.)* I was thinking of using nitrous oxide, unless you would prefer ether. It seems to me a little too much for a minor operation. I doubt whether I shall be four or five minutes in all. Just as you say, however.

VATABEEL *(With a dry, medical smile.)* Far be it from me to demand ether. I dislike the stuff intensely. *(He begins to take off his coat and waistcoat and adjusts an aseptic apron.)*

BAIL *(to Gailey)* I shall want a retractor, clamps and thumb forceps. Are all the different ligatures here? Ah, yes, I see. *(To Vatabeel)* Now, Doctor, if you will just make yourself comfortable. *(He indicates the operating table.)*

VATABEEL *(Opening the neck of his undershirt and sitting down on the edge of the operating table.)* I never imagined a small tumor could be so troublesome. *(To Bail)* This is where Greek meets Greek, isn't it?

BAIL *(When Gailey has unfastened the bandage around Vatabeel's neck, pressing the tumor lightly with his forefinger.)* But not bearing gifts unfortunately — at least, not pleasant ones. This seems to be doing very well; no inflammation.

VATABEEL *(Stretching himself comfortably, with, however, a sense of impending disaster or the possibility of it.)* At least this is the end of my bother with it.

The gas tank is wheeled forward, the breathing cap adjusted.

THE ANESTHETIST *(Taking his place at the doctor-patients head.)* Now, Doctor, if you please. We are only using one-fourth strength to begin with. And don't forget the forefinger.

VATABEEL *(Beginning to inhale and thinking of the mysteries of medicine and surgery and gases — to himself.)* Ah, yes, the forefinger. I must keep that going, or try to, until the gas overpowers me and I can no longer do it. When it drops of its own accord they will know I am unconscious. Marvelous progress medicine has made in these last few years! It hasn't been ten years since we had to administer ether and gas full strength because we didn't know how to dilute them. And there weren't any anesthetists. *(He begins to crook his finger.)*

THE ANESTHETIST *(One finger on Vatabeel's pulse, the other on the siphon regulator.)* That's very nice, Doctor, excellent. Breathe very deeply, please — as deep as possible.

VATABEEL *(Continuing his thoughts, but taking a deep, full breath.)* How self-contained and executive these young beginners are — just as I was in my day! Thus the control of the world passes from generation to generation. *(His face and ears begin to tingle. The fumes of the gas reach his brain. A warm, delightful stupor overcomes him. He imagines he is moving his forefinger, but he is not.)*

VATABEEL *(Noting the change.)* Very full breath, Doctor, if you please. Keep the finger moving as long as you are conscious. *(The finger moves feebly once or twice; then ceases. The arms and legs become inert.)*

THE RHYTHM OF THE UNIVERSE Om! Om! Om! Om! Om! Om! Om! Om! Om! Om! Om! Om! Om! Om! Om! Om! Om!

VATABEEL *(Functioning through the spirit only, conscious of tremendous speed, tremendous space, and figures gathered around him in the gloom.)* Strange! Wonderful! Astounding! This is the same place I was in when

I was operated on before. These are the same people. I hear voices. A most impressive company! *(The figures begin to converse.)* This is immensity — all space — that surrounds me. I am not alive, really, and yet I am. Am I so important as this? How dark, and yet how strangely light! *(Feels a sense of great heaviness and great speed.)* This operating table is moving like lightning! Who are these people about me, not Bail or Gailey? *(He thinks to see, but cannot.)* This is something else. I wonder if I shall come out of this! Oh, the terror! I really don't want to die! I can't! There are so many things I want to do. People do die under the influence of gas.

The arc of his flight bisects the first of a series of astral planes.

ALCEPHORAN *(A power of physics without form or substance, generating and superimposing ideas without let or hindrance. They come without word or form and take possession as a mood and as understanding without thought.)* Deep, deep and involute are the ways and the substance of things. Oh, endless reaches! Oh, endless order! Oh, endless disorder! Death without life! Life without death! A sinking! A rising! An endless sinking! An endless rising!

THE RHYTHM OF THE UNIVERSE Om! Om! Om! Om! Om! Om!

BAIL *(Turning from the examination of the instruments and examining the eyes of Vatabeel, turning the lids up; to himself)* A remarkable man, very. Such sacrifices for his profession! How persistently he has scorned money. Great, and poor — that is my idea of a physician. *(To the anesthetist)* How is he now, Doctor?

DRYDEN *(Who is holding Vatabeel's left wrist.)* Very good, I think. *(He looks at Gailey for confirmation.)* His pulse is one hundred and ten. His blood pressure seventy.

GAILEY He is quite under.

BAIL *(Lifting an arm and dropping it.)* Excellent! *(To Gailey and Tufts)* Turn him on his right side, please. The scalpel and the retractor, please. *(He takes up a scalpel and makes an incision one and one-half inches long by one-half inch deep. Tufts sponges the blood.)*

VATABEEL *(An inert mass carried in the line of the earth's arc and becoming conscious of it, but unconscious of pain.)* Oh, wonderful, wonderful! They are talking! It is light! It is dark! What is that they are saying? This rhythmic beat is so strange!

The arc of the earth bisects a second plane.

THE RHYTHM OF THE UNIVERSE Om! Om! Om! Om! Om! Om! Om! Om ! Om ! Om! Om! Om! Om! Om! Om! Om! Om!

FIRST SHADOW *(Of the second astral plane; a tall, grave man, seemingly with heavy dark whiskers and hair and deep blue eyes, surveying Vatabeel's body as it speeds onward and he with it.)* This man is of the greatest import, scientifically speaking, to his day. His trouble relates to Valerian, an element inimical to him. It is more serious than he thinks. It may be that he will not live. It may be that Valerian is unalterably opposed to him.

The voice becomes confused with other voices. Shadows gather about as though in conference. The operating table sweeps on at limitless speed.

THE RHYTHM OF THE UNIVERSE Om! Om! Om! Om!

SECOND SHADOW *(Seemingly near; a surgeon in contact with the wound.)* Very serious! Very serious! It lies closer to the large artery than they think. In fact, it surrounds it. A separating shield may help. This man should not be permitted to end yet. He is of great import to life.

Other figures gather about in the gloom and confer. The shadow increases. The voices cease.

ALCEPHORAN No high, no low! No low, no high! Time without measure, measure without time. A rising, a sinking! An endless rising, and an endless sinking!

VATABEEL *(Experiencing a vast depression as of endless space and unutterable loneliness.)* Ah!!!

THE RHYTHM OF THE UNIVERSE Om! Om! Om! Om!

The earth sweeps onward in its arc, bisecting a third plane.

BAIL *(Inserting a surgical spoon and scraping out the wound.)* This thing is somewhat more serious than I thought. I believe the tumor surrounds the large artery. It has ramifications I hadn't thought were here. *(To Gailey and one of the Nurses, bending over)* Here are two side pockets to the left and one just below. And another! We'll have to tie up some of these veins before I can go any farther. This artery is abnormally near the surface, to begin with. How is his pulse? (*He talks as he works, holding a bit of tissue up to the light, catching vein ends with hemostats, while Gailey ties the knots with silk thread and the Nurses pass thread and sponges.)*

DRYDEN *(In charge of the tank and feeding cone.)* One hundred and ten.

BAIL *(To himself.)* Excellent.

VATABEEL *(Sensing the line of the arc of his flight to be upward as yet.)* Strange, I feel so comfortable, yet so helpless — Jason James Vatabeel, physician extraordinary, scientist. Of so much importance. Will I live? Will I die? Life is so treacherous, so sad!

FIRST SHADOW *(Central figure of a new group, and a surgeon as the operating table rushes into a new realm)* Difficult! Difficult! This man is in a very serious condition much more serious than he imagines. The envy of elements! His services to life are in great danger. I am not sure that he can return to the world. (*He shakes his head with grave, oppressive solemnity, while the other shadows seem to listen and articulate.*)

THE RHYTHM OF THE UNIVERSE Om! Om! Om! Om! Om! Om!

ALCEPHORAN Deep below deep! High above high! No high! No low! Space beyond space! Singleness without unity! Unity without singleness!

VATABEEL *(Awed and disturbed by the rush and confusion.)* Spirits of the first order of earthly council. This mystery of living, how I have pondered it! Vast orders and powers of which I know nothing. The terror of the after life — what may it be — Death? Annihilation? No continuance? — Forever and ever? And in life itself — the mystery of the blood, of articulated bones, of organized society. Poverty, waste, hunger, pain, wealth, sickness, health — I have tried to think there was some good in what I've done. Vanity, hate, love, greed, patience, generosity. My fame is so wide, I know so little. *(He sighs deeply.)* Ah!

GAILEY *(Noting the tendency toward greater vitality, and so toward consciousness.)* A little more gas, perhaps. This cutting is affecting him.

DRYDEN *(Administering more.)* I think so.

BAIL *(Gouging at a second sac.)* This is apt to shake him a little. Perhaps ether would have been better, after all. It is going to take longer than I thought. How is your oxygen? *(He is thinking of how much gas will have to be administered and how much oxygen may be required to restore the patient.)*

DRYDEN *(Who has received his supply from the institution.)* All right, I think. *(He tries it. The examination proves that it is dangerously low. To Williams, his assistant.)* See if you can find another tank.

Bail frowns slightly, unconsciously irritated by the unpreparedness.

SECOND SHADOW *(Of the second group — a stern, almost invisible figure.)* I perceive near the cardiac region a tendency to weakness which is affected by gas. His condition is serious. Powers inimical and above us are at this moment producing error. This man is a powerful thinker and original investigator. Of him much might be expected.

The operating table sweeps on. The Rhythm of the Universe asserts Itself.

VATABEEL *(In vast depression, lying as under an immense, suffocating weight.)* Precarious! Precarious! And I do not want to die. I have so much to live for, so much fame to seek, so much to do. *(He sighs again.)*

DEMYAPHON *(Nitrous oxide, also with the power of generating and superimposing huge ideas without let or hindrance, the capacity of the individual permitting. They come without word or form, taking possession as a mood or as understanding without thought.)* So life is to be studied, and what for? Your little experiments! What do they teach you? You seek to find out, to know!

THE RHYTHM OF THE UNIVERSE Om! Om! Om ! Om!

ALCEPHORAN *(At an angle to the waves of Demyaphon.)* Vast! Vast! Vast! Measure without time — time without measure!

DRYDEN *(Noting Vatabeel's pulse to be greatly depressed and shutting off the gas, at the same time turning on the remaining oxygen.)* My assistant is long about that oxygen. See if you can find him, Miss Karns.

A nurse departs hurriedly.

BAIL *(Realizing that he has a much more treacherous situation at hand than he had imagined and anxious for the security of his patient; to the anesthetist.)* Don't let him get too low, Doctor. It is these extra pockets. I shall be done shortly. *(He hastens his efforts.)*

DRYDEN *(Becoming disturbed over the delay of the oxygen, and lifting an eyelid to observe the condition of the patient's eyes.)* Hm! I don't like the look of that. *(Aloud)* Chafe his feet, Miss Hale. You had better move his arm up and down. *(The oxygen gives out.)* I don't understand this oxygen business.

MISS KARNS *(Returning)* They have allowed the storeroom on this floor to run out. He has gone to the basement in the next building.

DRYDEN *(Snapping his teeth)* Run and tell him to hurry — please. I am all out. *(She departs.)*

DEMYAPHON *(Appearing only as thoughts placed in the dreamer's mind.)* There is a solution, but you will never be able to guess it. It is ages beyond a growth, which, when it is passed, you will be unable to remember. Eons upon eons, worlds upon worlds. Far and above the mysteries here and below are other mysteries — deep, deep. You puzzle over the phenomena of man. In a vain, critical, cynical ambitious way you dream. It will all be wiped out and forgotten. To that which you seek there is no solution. A tool, a machine, you spin and spin on a given course through new worlds and old. Vain, vain! For you there is no great end. *(A sense of ruthless indifference, inutility, futility, overcomes the spirit of Vatabeel.)*

THE RHYTHM OF THE UNIVERSE Om! Om! Om! Om!

ALCEPHORAN Behind, before, beneath, above, presence without reality — reality without presence.

FIRST SHADOW *(Of a third group, vague yet clear, young, experimental, curious, indifferent, obviously operating as a surgeon in charge.)* We shall soon be done with this now. He bleeds a lot, doesn't he? A bony old duffer! A ligature, please. A hemostat. I don't see why I should have been given this to do. They say he is needed. *(He seems to bend over. Other faces are near.)*

SECOND SHADOW *(Seemingly operating in charge of the gas and nose cone.)* It looks as though this gas might prove too much, Doctor. His pulse is a little feeble.

FIRST SHADOW *(Indifferently)* That can't be. We are two periods this side the danger mark on this plane. We can leave him until he reaches the next one. He's safe enough.

The operating table, like a bier, rushes on. The shadows recede. Once more darkness and space, and a sense of rigidity and tomb-like confinement.

THE ANESTHETIST *(Anxiously)* Will you please go and see what's keeping them, Miss Hale? He can't stand this much longer. His pulse is one hundred and forty now.

Miss Hale dashes from the room. Bail, conscious of the lapse of oxygen gas, increases his efforts to clean and close the wound.

THE RHYTHM OF THE UNIVERSE Om! Om! Om! Om!

DEMYAPHON *(Continuing)* So complicated that even the littlest things concerning man you cannot suspect. You think of forces as immense, silent, conglomerate, without thought, humor or individuality. I am a force without dimension or form, yet I am an individuality, and I smile. *(A sense of something — vast and formless — cynically smiling comes over Vatabeel, though he cannot conceive how. He is conscious of a desire to smile also, though in a hopelessly mechanical way.)* I am laughing gas, for one thing. You will laugh with me, because of me, shortly. You will not be able to help yourself. You are a mere machine run by forces which you cannot understand. This life that you seek you may have it on condition, by a condition. You will find out what that is a little later, yet you will not know for certain.

THE RHYTHM OF THE UNIVERSE Om! Om! Om! Om!

FIRST SHADOW *(Of a fourth group — a young doctor — material, much more material than the last.)* A little Valerian, please. Some iodine. Doing very well, don't you think, Doctor?

SECOND SHADOW *(In charge of gas and feeder cone.)* I am not so sure, Doctor. You will have to hurry. He isn't very strong. He should have been taken care of on the last plane. His eyelids —

FIRST SHADOW *(Working briskly but indifferently)* Nonsense! That can't be. He's one point this side the danger mark on this plane. No hurry. He'll do well enough.

THE RHYTHM OF THE UNIVERSE Om! Om! Om! Om!

THIRD SHADOW *(A nurse suggestive of mild materiality, bending over.)* He's sinking, Doctor, I tell you. He can't go much longer. Look at his hands! Look at them! He ought to be hurried to the earth plane.

The bier rushes on into space. The voices fade and cease.

THE RHYTHM OF THE UNIVERSE *(Resuming)* Om! Om! Om! Om! Om! Om! Om! Om!

ALCEPHORAN A rising, a sinking! An endless rising! An endless sinking! Outward without inward — inward without outward…

DEMYAPHON Material planes that recede — each one more material than the other, as you sink to your own. Spirits almost more material than yourself. Because of the points spoken of as in your favor, you think you will regain life. You do not know that they are standards set by you in previous experiences, eons apart. To live you will have to attain to a new one now.

VATABEEL Ah!

DEMYAPHON Round and round, operation upon operation, world upon world, hither and yon, so you come and go. The same difficulty, the same operation, ages and worlds apart. Your whole life repeated detail by detail except for slight changes. Now if you live you must make an effort or die. *(The gas smiles.)*

THE RHYTHM OF THE UNIVERSE Om! Om! Om! Om!

DRYDEN *(To Gailey)* He can't get back, Doctor, unless we get the oxygen here in thirty seconds. This tank is run out. His condition is desperate. *(He does the nurse's work, chafing one of Vatabeel's hands; to himself):* If he does, it will be the most wonderful case I ever heard of. A new standard, by George. *(He wipes the perspiration from his brow.)*

VATABEEL *(Struggling desperately to assist himself to live.)* A thing of the spirit, this, plainly. I suppose I am a test but how futile so to be. Round and round and round, an endless, pointless existence. Yet I cannot help myself. I must live. I must try. I do not want to die. *(He makes a great effort, concentrating his strength on the thought of life.)* Oh how ruthless and indifferent it all is. Think of our being mere machines to be used by others! *(He struggles again, without physically stirring.)*

MISS KARNS AND MISS HALE *(Hurrying in.)* Here it comes now!

THE ASSISTANT *(In charge of tanks, following.)* I had to go to the second building for the key. The floor man was over there. *(He quickly couples the connections and the oxygen is turned on.)* Fine work, I call that!

DRYDEN *(Bitterly)* What a system! And half a dozen important operations on today! *(He adjusts the cap and feeds the oxygen, full force.)*

VATABEEL There is something vastly mysterious about this — horrible! In older worlds I have been, worlds like this. I have done this same thing. Society has done all the things it has done over and over. We manufacture toys — the same toys over and over. Does Life produce its worlds and evolutions the same way? Great God!

DEMYAPHON *(Cynically)* The resistance which you are now displaying is in part by reason of your previous efforts and previous successes. You are the victim of experiences of which you have been made the victim. A patient, a subject, a tool, a method, round and round and round you go, a servant of higher forces, each time seemingly a step farther, each time in this way, for the same purpose, the same people, to no known end, over and over.

VATABEEL Ah!

THE RHYTHM OF THE UNIVERSE Om! Om! Om! Om!

GAILEY *(Disturbed by his weakened condition, and uncertain whether or not he can be revived.)* I am afraid that you will have to hurry, Doctor. He is very weak. His pulse is scarcely distinguishable.

BAIL *(Desperately scraping the last pocket and tying the veins.)* I am not supposed to be handicapped by poor service in this institution. Try to hold him a few moments. *(To Tufts)* Sponge! *(To the first nurse)* Scissors!

FIRST SHADOW *(Of a fourth group just outside the gates of life, a very material young doctor, whose hands and white uniform are almost luminous.)* Say, there isn't so much to do here — is there? A few stitches. Those veins ought to be clamped, though. *(He works briskly, lightly, with an inconsequential air.)* He'll do all right, don't you think?

SECOND SHADOW *(At the gas tank.)* Pretty weak, I should say. Gad, yes! He may hold out, though. They didn't shut off the gas on the last plane — that's a good sign. They usually do if it's serious. He's just at the turning point.

THIRD SHADOW *(A nurse apparently impressed by the uncertainty of the occasion.)* He's very low, I tell you, Doctor. Look at his nails. You'd better shut off the gas. He's nearly all in! Look at his eyes! He's william, I tell you. He's william. He can't live thirty seconds more.

An intense, disturbed rate of vibration indicates crisis. The second shadow shuts off the gas. The operating table rushes on into darkness.

VATABEEL *(Thinking)* On, on — and I am now to die — I am dying, unless I can help myself! An endlessly serviceable victim — an avatar! The mystery of life — its gloomy complications! But I don't want to die! I won't die. *(He concentrates vigorously on the thought of life.)*

DEMYAPHON *(Smiling)* The points which you established on your previous circuit of this orbit of materiality, and which have been counting in your favor, have now been exhausted. This safety mark, which you have heard frequently mentioned, you yourself established. If you live it will be by setting a new standard — rendering a new service but in an old way — over and over and over. Unless you struggle to live — unless you succeed in living —

VARIOUS VOICES Try, oh try, oh try! You, above all others!

Vatabeel senses some vast, generic, undecipherable human need. He wishes to weep, but cannot.

THE RHYTHM OF THE UNIVERSE Om! Om! Om ! Om!

A sense of derision, of indifference, of universal terror and futility, fills Vatabeel. Suffocating, he tries to move.

ALCEPHORAN Deep below deep! High above high! no beginning, no end! No end — no beginning!

VATABEEL *(Terrified and yet seemingly helpless.)* The dark! The dark! The ultimate dark! Plane upon plane! Eon upon eon! To do over and over! Or annihilation! Why — oh — why. But I won't die. I can't. *(He struggles again.)*

DEMYAPHON It has no meaning! Over and over! Round and round! The orbit of which you are a part brings you back and back again and again in non-understanding. *(The thought seems to become rhythmic and painful.)*

VATABEEL *(Struggling)* Am I really to die? Oh, no!! What if I do go round and round! I am a man! Life is sweet, intense, perfect! If I do go round and round, what of it? Beyond this, what? Nothing! I serve! *(He stirs. His spirit struggles with materiality. The vital spark is rekindled within the inert frame. With a gigantic effort, it re-establishes itself and resumes control and respiration. The effort to inhale, feeble at the surface of materiality, is immense.)*

DRYDEN *(Working the one free arm as rigorously as possible, while Miss Hale and Miss Karns chafe his hands and feet.)* There, he has caught it. Chafe his arms, Miss Karns. I am not sure that we can bring him round even yet. His vitality is amazing. I don't understand it at all. His heart was all right, though — extra strong.

BAIL I shall have to have a few more seconds. I have three stitches to take. You may let him come out if you wish. This is the last time I shall use gas — here. I have had enough trouble with it before. *(He tries to think where.)*

DEMYAPHON *(To Vatabeel)* And the humor of it is that it is without rhyme or reason. Over and over! Eon after eon! What you do now, you will do again. And there is no explanation. You are so eager to live — to do it again. Do you not see the humor of that?

With sardonic intent the rate of vibration which is laughter is set up in Vatabeel's body. Even as he struggles to breathe and to regain his material state, he realizes that the impulse, a part of something vast, unearthly, mechanical, wavelike, is sweeping him into its rate. Weak from loss of blood — in danger of rupturing the large artery in the centre of the wound, close to the surface, he begins to swell with pent-up laughter. A dry, hard, sardonic desire to shout overcomes him, although he is yet unable to move.

DRYDEN *(To Gailey, noting the customary action of nitrous oxide as the patient approaches consciousness, and uncertain what to do.)* He is coming to. I'm a little afraid to use more gas in his present condition, Doctor. If he laughs too hard — !!

BAIL *(Irritably)* Can you keep him under ten more seconds? I have one more stitch to take. *(He takes one.)*

DRYDEN I think he'll last that long, Doctor, anyhow.

Nurses and assistants seek to hold Vatabeel rigid in order that the operation may not be disturbed.

DEMYAPHON And I told you you would laugh. You will eventually forget why, but you will shout and shout and see no reason. I am the reason. I am the master of your personality. I am Demyaphon — Laughing Gas. Shout! Shout! *(It leaves him in a waking condition.)*

VATABEEL *(As Bail takes the last stitch and Gailey begins the bandaging of his neck, seemingly bursting into consciousness, the wound still unbandaged, the pain of the needle still fresh.)* Oh, ho! ho! ho! Oh, ha! ha! ha! Oh, ho! ho! ho! Oh, ha! ha! ha! Oh, ho! ho! ho!

GAILEY *(Holding one arm to calm him, uncertain as to whether he is mentally clear or not — as yet.)* Something very funny, Doctor?

BAIL *(Accustomed to the effects of laughing gas, but disturbed by his patient's condition — to Gailey.)* Make those bandages very tight. I'm afraid of that wound. It is too bad we couldn't have kept him under longer. He's very close to death even yet. I scarcely had time to take those stitches properly. And, of course, the effects of the gas have to be the very worst possible. *(He shrugs his shoulders.)*

VATABEEL *(Still shaken by the rate of vibration set up in him, his mouth open, his face a mask of sardonic inanity.)* Oh, ho! ho! ho — oh, ha! ha! ha! Oh, ho! ho! ho — oh, ha! ha! ha! Oh, ho! ho! ho — oh, ha! ha! ha! I see it all now! Oh, what a joke! Oh, what a trick! Over and over! And I can't help myself! Oh, ho! ho! ho! Oh, ha! ha! ha! And the very laughing compulsory! vibratory! a universal scheme of laughing! Oh, ho! ho! ho! Ah, ha! ha! ha! I have the answer! I see the trick. The folly of medicine! The folly of life! Oh, ho! ho! ho! Oh, ha! ha! ha! Oh, ho! ho! ho! Oh, ha! ha! ha! What fools and tools we are! What pawns! What numbskulls! Oh, ho! ho! ho! Ah, ha! ha! ha! *(His face has a sickly flatness, the while he glares with half-glazed eyes, and shakes his head.)*

GAILEY I never saw gas act more vigorously. Did you, Doctor?

BAIL *(Annoyed by the incident.)* I never did. *(Taking his friend's arm.)* Come, Jason, you're all right now! Get over this! Just laughing gas, you know. It's all over. You have a serious cut in your neck. *(He presses his arm fondly.)* You're just laughing because of the gas.

VATABEEL *(Wearily — with the sense of immense futility still holding him.)* Oh, ho! ho! ho. Oh, yes, yes, yes. Just laughing gas! And that's why I laugh. Oh, ho! ho! ho! Ah, ha! ha! ha! I don't wonder it laughs! I would too! You would if you knew! The mystery! The cruelty! The folly! Oh, ho! ho! ho! Oh, ha! ha! ha! *(He stares and glares the while his friends and hearers view him with kindly, condescending tolerance mingled with a touch of awe and amazement.)*

BAIL *(Genially)* Just the same, it's all over, Jason. Come on!

VATABEEL *(Shaking himself and beginning to recover his natural poise and reserve.)* And was it only the gas, then? That is very strange: I thought — I thought — I wonder? I wonder — ? *(His mouth remains open.)*

DRYDEN *(His calmness restored)* It seems odd to see him laughing like that.

GAILEY The fumes are still in his head. He'll be all right now, though. That was a pretty close shave. I thought we had lost him. There'll be a new storekeeper here tomorrow, if I have my way.

THE SECOND ASSISTANT I never saw Dr. Bail so irritated. He'll hold this against us.

The various doctors and nurses and assistants go about their duties. Bail slowly leads Vatabeel to his automobile. Vatabeel's face retains a look of deep, amazed abstraction.

CURTAIN

12

"I took my station on a base of infinite nothingness"

"The Chair of Metaphysics", an article appearing in the September 1920 edition of *The Atlantic Monthly*, in which the anonymous author tells of the philosophical opportunities afforded by tooth extraction.

THE CHAIR OF METAPHYSICS

The one requirement I make of dentists is that they shall be able to administer the consolations of philosophy during the prosecution of their researches in physical torture. A dentist may have all knowledge and all charity, together with all the latest textbooks and a surgeon's case full of the most perfected and lethal implements of his trade; he may have a resounding reputation and a clientele of the most distinguished impressiveness; but if he have not philosophy, all the rest is as nothing.

On the other hand, let him beguile the hours of torture with impersonal discourse, tending from my immediate pain to the secrets of anatomy, and thence through psychology to cosmic themes, and I forgive him everything; nay, I remember my hours with him as among the pleasant ones of a variegated personal history, and value his ministrations above those of the more knowing efficient ones who, I realize, are later to criticise his technique, damn his methods, and undo his work, as a preparation for doing it over in their own more expensive ways. For to me the chair of dental torture has always been, essentially and inherently, the chair of metaphysics.

There must be many who find that the body's pain unbars curious doors of speculation, admitting the mind to broad halls and chambers of impersonal thought which, in the hours of normal comfort, remain unvisited. Pain is an elemental thing; it knocks and pries and twists at the very root of individual consciousness, and strikes so much deeper into the mysteries of being than anything else does, that it forces one to the brink of startling discoveries about space, time, and the ultimate secrets of things. Extreme pain forces us to the wall, the limit of the personally endurable; and we grope for the one possible way out, which is the way of impersonal thought. Many, I say, must have learned this fact — by living it.

But I dare say there are not so many who, like myself, find that the door yields more readily when another helps one push it open, and who can fare farther along those strange dim cosmic highways companioned than alone. It is, if you like, paradoxical and perverse. Yet the fact remains that, at odd times, I have positively relished the nibbling of dental instruments at an exposed nerve, and used it as the point of departure for the airiest philosophic flights — granted only a practitioner who spoke my language and was always ready with a theory for my facts or a criticism for my theory. On the other hand, with a dentist whose mind was only on his work, I have found in a mere painless cleaning of the teeth the most prosaic and unredeemed torture, and cursed each crawling minute, together with the dial that recorded it. Give me, every time, the physician who is also a metaphysician.

My first dentist I may be said to have inherited. At least, as a young man he had attended my grandfather, and in middle life my father and my uncle. Then, as a sallow, dignified, rather frail old man, he attended me — just once, when, a gawky lad of seventeen, I went to him because a certain molar and I had come to the parting of our ways. He was a stiffly dignified old gentleman, in both mind and outward appearance. His mental operations were chiefly Calvinist, strongly and very oddly tinctured with science.

I went to him in vacation, and found him in his dingy office at the top of a three-story "block" in the little home town. He received my name with a patient abstractedness. It was manifest that he had forgotten me. He was, in fact, so aged that he was fast forgetting everything merely personal.

He seated me in the operating-chair facing the two windows above the street, prepared the tank of gas, and began slowly administering it. Remembering my last previous experience of an anaesthetic and my writhing struggles with black and demoniacal shapes, I had resolved that this time I would be the most tractable of patients. Accordingly I gripped the arms of the chair and braced my whole body rigidly, in the intense effort of self-command. Presently I heard the gentle voice of Doctor Zachary, heaving toward me in hollow waves from across a black void: "There now, there now; relax, just relax." I relaxed; and further, to show him the docility of my compliance, patted the arms of the chair in a soothing and reassuring gesture, as if to say, "Have no fear of me, my good sir!"

To my astonishment, I could by no means stop doing this, once I had begun. I went on increasing the amplitude and the velocity of my gesticulations, until I was swinging my arms through the air like great flails, beating him away from me, striking down his apparatus, and behaving like an insane demon generally — all in pure goodnature gone daft, like that of the trained bear who crushed his sleeping master's skull in killing the fly on his master's forehead.

We began over. This time I just relaxed, without any urbane attempt to demonstrate the completeness of my relaxation. And this time the gas accomplished what was desired of it, or at least a part thereof. I drifted out among star-ways, and a galaxy of saffron constellations whirled about my head. In some outer void of space I took my station on a base of infinite nothingness. Presently a great yellow world sped by me with the speed of a cannon-ball, yet deliberately enough so that I could read, in characters of flame, against its black equatorial belt, the figure 1,000,000. Another world sped after it in the same track, similarly inscribed, but with a figure of still huger magnitude. After this one came another; then others still, the series increasing in size and speed, and each world marked with a figure greater than any that had appeared before, until they had mounted to sums for which there is no designation in human speech. And as they receded and fled away, diminishing down the appalling void, and I felt myself blown upon by the cosmic winds of their passage, it was borne in upon me somehow that I was now contemplating the cycles, not of time, but of mortal sin. These that sped by me were the aeons upon aeons of sin through which the stellar universe must win to its ultimate purification. Eventually there was to come, in the wake of all, a world white and lucent, gleaming like the plumage of an angel's wing. It would mean that the planetary system had won through turmoiling cycles of sin to its redemption. There rang in my ears an immemorial phrase, "The Blood of the Lamb," and a surge of cosmic music, somehow crimson, was to engulf me.

But alas! this white and splendid consummation depended on my fully taking in, with my one poor finite unarithmetical brain, each and every one of those staggering figures. If I missed so much as a single cipher, the whole universe was lost to darkness and dissolution, it might be for ever and ever.

Suddenly I had come against a purple veil — the uttermost firmament of all things that ever were. Only, a part of me seemed to be on one side of the veil, and a part on the other. The part on the other side was the supernal part which, if I could but get at it, flow into it, could take in those colossal figures, and by comprehending them save the universe from lapsing into aboriginal chaos, perhaps to begin its weary cycle all over. Somehow I must make the two sundered parts of myself fuse and coalesce — the part which was mortal and finite and baffled, and the liberated and untrammeled part beyond. To accomplish this there was no way but to rend the veil. When I had done this, I should comprehend within myself all that has ever been, is, or shall be.

With a convulsive and superhuman effort I laid hold of the veil with both hands and strove to tear it. At the same time something seemed to tear madly at the fabric of my own being. A dull explosion shattered the universe about me — only, in some curious sense not open to definition, it seemed to be the inside of my own head that had exploded. There came a sudden silvery tinkle of music, incommunicably sweet, and —

I sat bolt upright in the dentist's chair. An unusually tall person, I had contrived to thrust a foot firmly through each of the pair of windows in front of me, and from the street below came the last tinkles of the falling splinters of glass. In my two hands I still clutched the two halves of Doctor Zachary's glossy Prince Albert coat, which in my final paroxysm I had seized by the tails, reaching behind the old gentleman, and split clean up the back by the simple act of spreading my arms apart as far as they would go.

Doctor Zachary stood there panting and disheveled, but not abating by one jot the mild dignity of his usual air. In the ruin I had created, the first aspect to come uppermost in his mind was the foolish circumstance that his forceps, on which he had not for an instant remitted his grip, had miraculously come to light at the wrong end of his sleeve.

The nerve of the misbehaving molar, barely loosened from its hold on my lower jaw, jumped in a savage rhythm, as an infuriated beast springs to the length of its chain over and over, or as your heart pounds in a sudden deadly fright. I was maddened with the pain — but I was more maddened by the interruption of my all-but-consummated dream. The denouement was unspeakably prosaic. But I was unspeakably above prose.

Doctor Zachary reasphyxiated me, and the extraction was achieved in short order. When I was conscious again, he quizzed me, not without an effort of sympathetic understanding, about the sensations which had dictated my grotesque behavior. And as I went on groping for the words to recreate my cosmic vision, he began to nod more and more frequently. And then, for something like an hour, we discoursed together on the various philosophies of sin and judgment, coming round in the end to the problem whether there can be, modernly, any such thing as direct revelation.

I see now that many of his ideas were old-fashioned, out-moded; if I heard them restated, I should doubtless be filled with abhorrence. But to a boy just beginning the painful processes of thought, it was a famous dialogue — the first halting parasang of an intellectual anabasis which may perhaps be cut off when this machine stops, but which I trust is never really to be completed. Anyway, this talk laid the foundation of my permanent requirement that the dentist's operating-room contain the chair of metaphysics.

When next I had need of Doctor Zachary, he was dead.

His successor in that same office, a profoundly reflective young man just out of dental school, carried me through several successive developments in metaphysical rationalism. He was withal a person of delightful tact. He knew how to make me laugh at my own grotesque attempts to converse on lofty themes through, or round, a rubber dam, without in the least degree laughing at them himself; though the language I spoke must have been practically devoid of consonants, and in fact more than a little like that of a man remembered from my early childhood, who was whispered to be without a palate and almost without a tongue. His ideas touched the central problems of consciousness exactly when his burr touched the central nerves of the tooth. Altogether he was a most gifted and indispensable young man; and never a session with him so painful that I could not have wished it longer. When he went to Colorado,— for a month or two,— I waited and let my teeth go to pieces for two years and a half, because I could not force myself to go to anybody else.

But he never came back.

There was a succession of others, of greater and less expertness professionally. I valued each according as it had or had not pleased his Maker to endow him with philosophy. Latterly, I parted with a wisdom tooth, as hard-earned (and as useless) as wisdom itself is often said to be, at the hands, or forceps, of a very modern, very efficient young dental surgeon of absolutely no capacity for generalization. He has — you know the sort — one of those utterly concrete minds.

He injected a local ansesthetic with the hypodermic needle. Then he stared out of the window and drummed with his knuckles on the sill for a minute, in impassive silence, waiting for the stuff to "take." Then, with a pretty little blued-steel knife, he slashed the gum here and there to make sure that there was no feeling left in it. There was indeed none. In its incapacity for sensation, it was precisely like his own mind.

His forceps closed on the tooth. He rocked it gently this way and that. Then there was a barely audible *spat!* of some hard particle falling upon crumpled paper. My contemptible wisdom tooth had somehow got into his wastebasket.

It was a triumphantly perfect job of its kind. Considered purely as a technological achievement, it was immense. The whole transaction, from my arrival at the office, could hardly have taken six minutes. It was as devoid of ugliness and pain as of philosophy. There could have been no greater contrast to the methods of old Doctor Zachary, locked with his patient in a prolonged and seemingly deadly struggle over the possession of one insignificant bit of white bone with a jumping hot pain at the centre of it. And yet —

Well, others may elect bare science and they will. But as for me, give me philosophy every time.

Index of Exclamations and Similes

EXCLAMATIONS

[26] "what an amazing concatenation of ideas"

[28] "Nothing exists but thoughts!—the universe is composed of impressions, ideas, pleasures and pains!"

[35] "TONES!"

[48] "let me breathe it again, it is highly pleasant! it is the strongest stimulant I ever felt!"

[78] "nitrous oxide"

[81] "Oh if such is heaven, then indeed it is desirable"

[84] "That tyrant! has seized my cane — deliver it to me! — this — instant! — or — I'll be the death of you!"

[84] "Byy hea-vens! — 'Twere nobly done! — To snatch the briidal honours — from the blaazing sunn!'"

[84] "Well, doctor, here I am, at last"

[85] "Alexaander!!!"

[85] "Lord! deliver us!"

SIMILES AND COMPARISONS

I felt...

[31] "like the sound of a harp"

[33] "to be bathed all over with a bucket full of good humour"

[35] "as if composed of finely vibrating strings"

[38] "[feelings which] resembled those produced by a representation of an heroic scene on the stage, or by reading a sublime passage in poetry when circumstances contribute to awaken the finest sympathies of the soul"

[45] "[feelings] resembling that which I remember once to have experienced after returning from a walk in the snow into a warm room"

[47] "as if all the muscles of the body were put into a violent vibratory motion"

[47] "as if I were lighter than the atmosphere, and as if I was going to mount to the top of the room"

[48] "[as if] ascending some high mountains in Glamorganshire"

[49] "[feelings] that I can compare ... to no others, except those which I felt (being a lover of music) about five years since in Westminster Abbey, in some of the grand chorusses in the Messiah, from the united powers of 700 instruments of music"

[50] "a tingling in the elbows not unlike the effect of a slight electric shock"

[77] "placed on an immense height, and the noise occasioned by the reiterated shouts of laughter and hallooing of the by-standers appeared to be far below me, and resembled the hum or buz which aeronauts describe as issuing from a large city, when they have ascended to a considerable height above it"

[78] "I was an inhabitant of the Elysium of Rousseau, or the island of Calypso, of Fenelon, blown by a rudely malicious blast into a world of reptiles where the atmosphere like the pestiferous samiel of the desarts of Arabia, was pregnant with destruction, and threatened inevitable annihilation to all who inhaled its morbid breath"

[85] "[feelings which] I could not but then compare the theatric rhapsodies to which I had been a witness, to the bombastic effusions of our western generals on entering Canada"

Gallery of Images

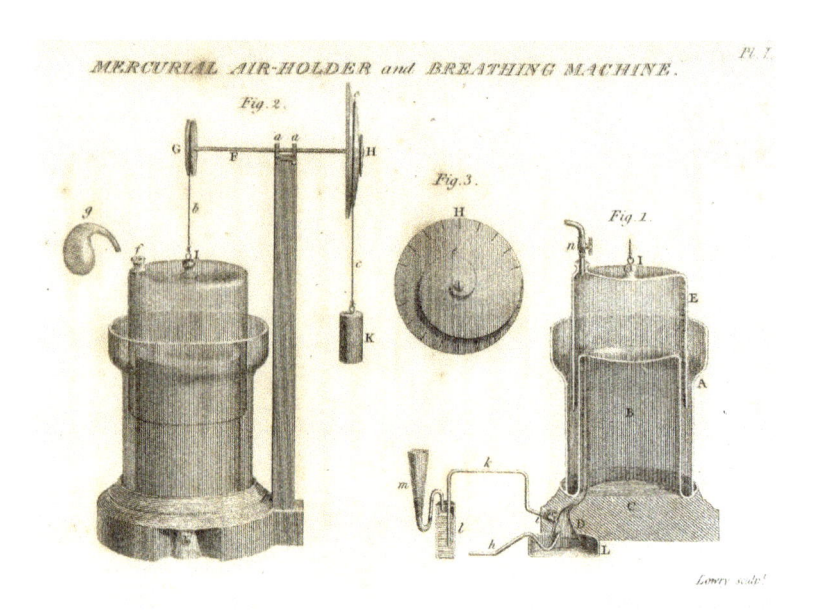

Illustration of apparatus used for the breathing of gases, frontispiece to Humphry Davy's *Researches, Chemical and Philosophical: Chiefly Concerning Nitrous Oxide, or Dephlogisticated Nitrous Air, and Its Respiration* (1800) — Source: Wellcome Library, London (CC-BY 4.0).

[128]

Scientific Researches! – New Discoveries in Pneumaticks! – or – an Experimental Lecture on the Powers of Air (1802), a satirical illustration by James Gilray. Leading member of the Royal Institution Sir John Hippisley is shown administering the experiment, while Humphry Davy hovers behind grinning and brandishing a pair of bellows. Gilray was a regular contributor to *The Anti-Jacobin Review*, the "ultra Tory" publication in which Richard Polwhele's satire *The Pneumatic Revellers* (p.55) was published — Source: Wellcome Library, London (CC-BY 4.0).

Dr Syntax and his Wife Making an Experiment in Pneumatics, a coloured aquatint by Thomas Rowlandson featured in *Dr Syntax in Paris; or a Tour in the Search of the Grotesque; being a Humorous Delineation of the Pleasures and Miseries of the French Metropolis* (1820), by William Combe — Source: Wellcome Library, London (CC-BY 4.0).

The Whimsical Effects of Nitrous Oxide Gas, frontispiece to John Joseph Griffin's *Chemical Recreations: A Series of Amusing and Instructive Experiments, which may be Performed Easily, Safely, and at Little Expense* (1825) — Source: Wellcome Library, London (CC-BY 4.0).

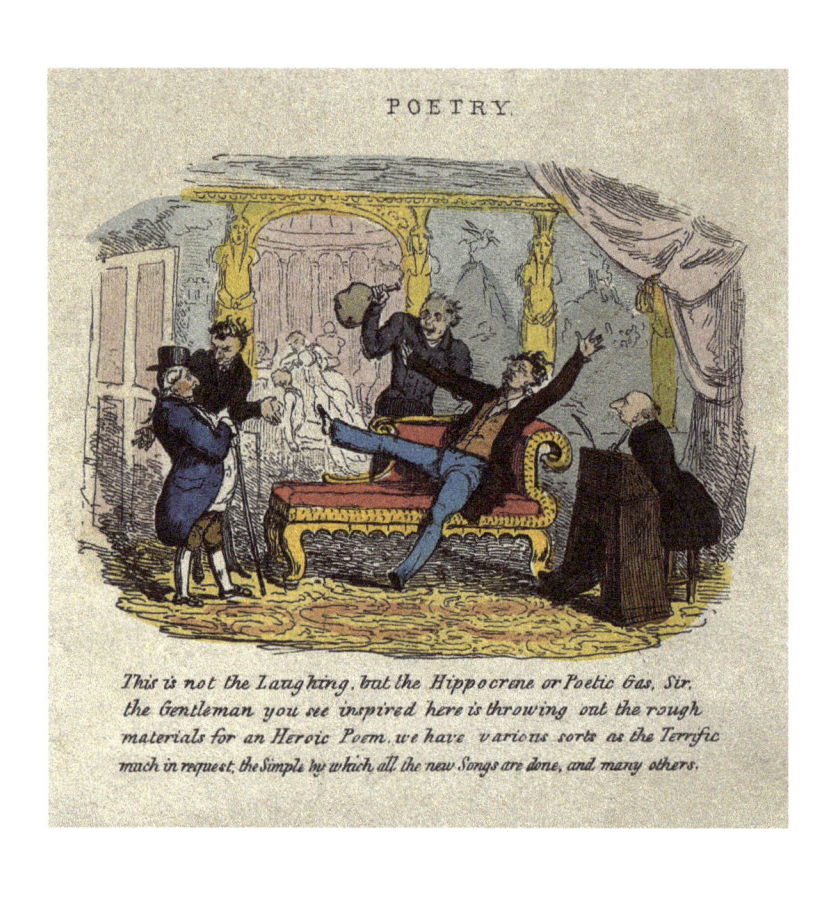

A group of poets carousing and composing verse under the influence of the "poetic gas", detail from a coloured etching by Robert Seymour, 1829 — Source: Wellcome Library, London (CC-BY 4.0).

Prescription For Scolding Wives, coloured etching from Robert Seymour's *Living Made Easy* series (1830). Humphry Davy is the central figure administering the dose of laughing gas while Count Rumford looks on from the left — Source: US National Library of Medicine.

Title page to the musical score of "Laughing Gas: A New Musical Comic Song" (ca. 1830), composed by W. H. Freeman — Source: Wellcome Library, London (CC-BY 4.0).

Laughing Gas.

"Laughing Gas" by George Cruikshank, the frontispiece to from John Scoffern's *Chemistry No Mystery* (1839) — Source: Wellcome Library, London (CC-BY 4.0).

EXHIBITION OF THE LAUGHING GAS.

The Nitrous Oxide, or Laughing Gas, was discovered by Dr. Priestly, who produced it by abstracting a part of the Oxygen from the Nitric Oxide. It is composed of equivalent parts of Oxygen and Nitrogen. Before the time of Sir Humphry Davy, it was considered irrespirable : but by some very interesting experiments, he proved this opinion to be incorrect ; he also wrote a work, entitled, "Researches on the Nitrous Oxide." It is named Laughing Gas on account of the very exhilarating emotions produced in those who respire it for a short time: laughing, dancing, jumping, acting, reciting, and (last but not least) fighting are amongst the prominent effects displayed by persons under its influence. The Febrile Miasma depresses and terrifies the mind as much as the Nitrous Oxide raises and enlivens it. The easiest way of making it is to dissolve Crystals of the Nitrate of Ammonia in a retort, over a strong flame ; after the atmospheric air has passed away, the Gas will be given off in great abundance, and may be collected in bladders, or a gasometer, for use. Sulphur, Phosphorus, red hot Charcoal, or a Taper, will burn with great brilliance when immersed in Nitrous Oxide.

Engraved and Printed at the Exhibition — H. & A. Hill, Printers, Castle Green, Bristol.

"Exhibition of the Laughing Gas", a wood-engraving from ca. 1840, most likely used as an informative poster at one of the many itinerant exhibitions showcasing the effects of laughing gas — Source: Wellcome Library, London (CC-BY 4.0).

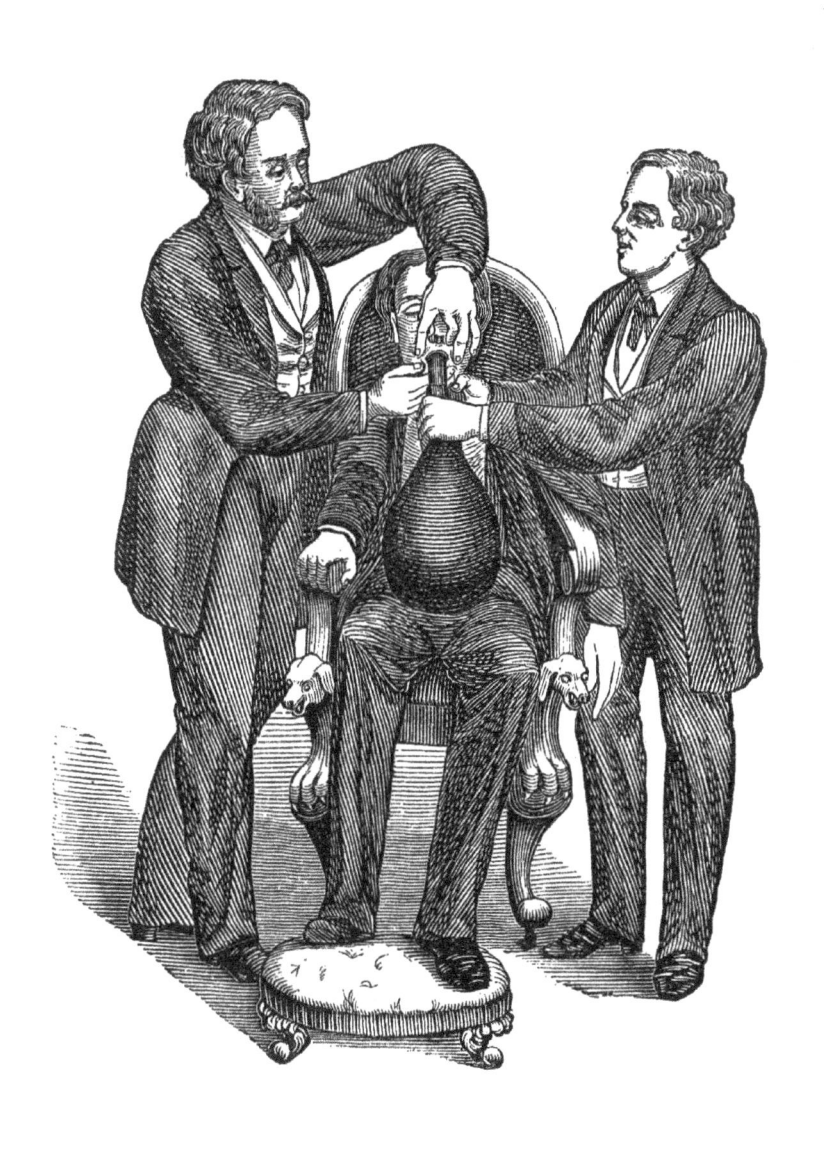

Illustration showing the surgical administration of nitrous oxide, featured in *Manual of the Discovery, Manufacture, and Administration of Nitrous Oxide, or Laughing Gas, in its Relations to Dental or Minor Surgical Operations* (1870), by Philadelphia-based dentist F. R. Thomas — Source: Wellcome Library, London (CC-BY 4.0).

DR. NEVIUS IN THE ACT OF ADMINISTERING NITROUS OXIDE GAS, COLTON DENTAL ASSOCIATION, COOPER INSTITUTE, NEW YORK.

Photograph of a nitrous oxide administration at the Colton Dental Association, from *The Discovery of Modern Anaesthesia* (1894), by L. W. Nevius — Source: US National Library of Medicine.

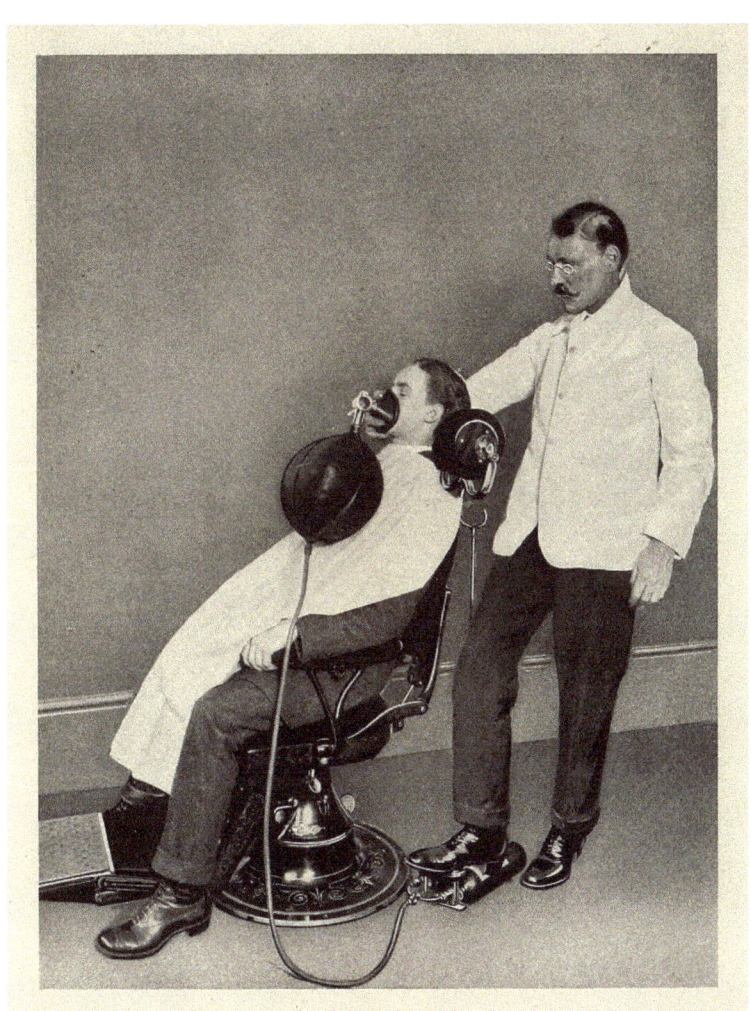

Pose of patient during the administration of nitrous oxide gas. The anæsthetist is placed behind in order to show the position of the patient and the face-piece. Normally he would use his left hand to hold the face-piece, not his right as shown in the figure, and he would stand on the left of the patient.

Demonstration of how an anaesthetist should seat their patient for the administering of nitrous oxide, featured in *Anaesthetics: Their Uses and Administration* (1914), by the influential London-based anaesthetist Dudley Wilmot Buxton — Source: University of Toronto via Internet Archive.

With thanks to *The Friends of The Public Domain Review* for their ongoing support:

A. P. Guttshall, Adele Fasick, Adrienne Anderson, Aemilia Scott, Afton Lorraine Woodward, Albert Jaccoma, Alé Mercado, Alembic Rare Books, Alexander Nirenberg, Alison Bell, Alison Wishart, Alissa Likavec, Allana Mayer, A. M. Emerson, Amanda Glassman, Amity Michelina, Amy Conger, Amy Harmon, Ana Méndez de Andés, Andrea Callow, Andrea Friedrich, Andrea Vela Alarcon, Andres Saenz De Sicilia, Andrew Chapman, Andrew Clifford Pinder, Andrew O'Kelly, Andrew Seeder, Andrew Wansley, Ann McCarthy, Ann Zatarain, Anne Carol, Anne Marie Houppert, Annette McKinnon, Annie Johns, Antonio Merenda, Anzan Hoshin, Armitage Shanks, ARTEL, Artist's Proof Editions, avianovum, Barbara Bradshaw, Barbara Ruef, Bea Hartman, Bella Terra Publishing, Ben Bradley, Bertrand Monier, Betsy Pohlman, Braden Pemberton, Bree Burns, Brenda Kimsey Warneka, Brian Floca, Brian Harmon, Bryan Gardiner, Burton Cromer, Buster T. Kallikak Jr, C. R. Crackel, Caitlin Keegan, Cally Wight, Candia McWilliam, Carl Taylor, Caroline Carlson, Caroline Evans, Cathy Sandifer, Cei Lambert, ChameleonJohn, Cheryl Powell, Chris Burns, Christian Fredrik Aronsen, Christian Jones, Christina Hammermeister, Christine A. Jones, Christine Kanownik, Christopher Berry Lethbridge, Christopher Hughes, Cigornia, Cindy Womack, Clara Bosak-Schroeder, Claudia Swan, Claudio Ruiz, clickscape.net, Colin Fanning, Corey Chimko, Cowbelles, Cristina Bryan, Curtis Thomas, D. P. Carroll, Dan MacDuff, Danae Panchaud, Daniel F, Daniel Lander, Daniela Didier, Darcie DeAngelo, David Bryan, David Chatwin, David Conolly, David Palmeter, David Sharpe, David Wolske, Debi Geroux, Deborah Weber, Deborah Woodman, DIAGRAM, Diane Fox, Diane Mayr, Dictionary of Sydney, Dina Eastwood, Dipesh Navsaria, Donna Emsel Schill, Dorothy A. Yule, Doug Harris, Doug Temkin, Dr Natalie McDonagh, Dr Richard O'Flynn, Dr Shuler Harmon, Dr David Abbey, Dr Lauren K. Robinson, Dr Omed, Dr Steven R. Miller, Dr C. Benedict, Dry Toasts, Dylan Flesch, E. Lee Eltzroth, E. A. Craig, Edinburgh Medical School, Elaine Barclay, Elettra Gorni, Elizabeth Novak, Elizabeth Rowe, Elizabeth Van Pelt, Elly Catelli, Elvira Piedra, Emily Forgot, Emily H. Cohen, Erica H. Smith, Erik Spiekermann, Erin Fletcher Singley, Erin McKean, Eugenia Leftwich, Far Beyond Film LLC, Fingal, Fiona Winifred Wood, Foxpath IND, Frances Gillis-Webber, Frank Kloos, Frank Modica, Futureofthebook, Gail Horvath, Geetartha Darshan Barua, Giampaolo Luparia, Gloria Katz Huyck, Greg Lehman, Grow House Grow, Guglielmo Centini, Guido Castellanos, Hannah Jenkins, Hannah Margolin, Harald Walter Azmann, Heather Hogan, Heavy Eyes, Henrietta Rose-Innes, Hilde Luytens, Ian Herbert, Ideum, Imogen Clarke, infoclio.ch (Swiss portal for the historical sciences), J. Bunning, Jack O'Connor, Jack Whoppy, Jackie Brooks, Jackie May, Jake DeBacher, Jakia, James Ashner, James Cox, James Downs, James West, Janie Geiser, Jaye Bartell, Jean Corkill, Jeanne Marie Neumann, Jeannine Jenkins, Jed Lackritz, Jeff Diver, Jeffrey Brian Hanington, Jeffrey Hamilton, Jeffrey Turco, Jemima McDonald, Jennifer A. Meagher, Jenny Burger and Jordan Carroll, Jenny McPhee, Jenny Molloy, Jenny Zigzag, Jerome Handler, Jesse L., Jesse U., Jesseca Ferguson, Jessie Huffaker, J. F. and Penelope Englert, Jill Littlewood, Jim Erickson, Joanne Koreman, Jockjimy, Jodie Robson, Joe Virbasius, Joel and Sarah Haffner, John Aboud, John Barrett, John Cooper, John Doba, John J. Griffiths, John Lopez, John Phillip Cooper, John R. Gibson, John Rooke, John Son, Jonathan Geer, Jonathan Gray, Jonathan Green, Jonathan Hirshon, Jonathan Lamb, Jordan Guzzardo,

Jose M. Diaz, Joseph Haley, Joseph McAlhany, Judith Field, Judy Hill, Julie Green, Julie Heller, Julie Trainor, K. Shafer, K. Sommerville, Karen Bonsignore, Karyn Kloumann, Kate Borowske, Kate MacDonald, Katharine Phenix, Katharine Pitt, Katherine Alexander, Katherine Grifin, Katherine Hall, Kathleen Sweeney, Kati Schardl, Katy Cherry, Katy Liljeholm, Keith Calder, Kenneth Whittaker, Kerrie and Simon Triggs, Keven Eyre, Kristen Gallerneaux, Kristie Mitchell, Kristine M. Richards, Kubi Ackerman, Lahela Nihipali, L. D. Gunther, L. E. Usher, Lang Thompson, Larry A. Schroeder, Laura Gibbs, Laura Griffiths, Laura Lindgren, Laura Lorson, Laura Macfehin, Laura Stein, Lauren Turner, Lawrence Wilkinson, LCTV, Leslie Gardner, Liam Guilar, Lieke Ploeger, LIGHTNER MUSEUM, Lilith Saintcrow, Lillian Wilkie, Lincoln County Television, Linda Bourke, Linda Rodriguez McRobbie, Lindsay van Niekerk, Lisa G O'Sullivan, Lisa Jackson, Liz Folk, Liza Daly, Lizzie Seal, L. M. Rima, Lois Blood Bennett, Lucia Mesak, Lujean Martin, Luke Sperduto, Luna Labrabeagle, Lydia Pyne, Lynn Rogan, Lynne Pate, Madeleine Stern, Maggie Simonelli, Magpie Bookshop, Malibu Carl, Marcella Vee, Marcia Moore, Marcus Ivarsson, Margaret E. Smith, Margery Meadow, Margherita Peliti, Maria Castello, Mariko Gordon, Marina Montanaro, Marina Nichols, Marjory Lehan, Mark Cohen, Mark David Kaufman, Mark H. Larick, Mark Stein, Markus Nietzke, Marly Gisser, Martha J. Fleischman, Martin L. Smith, Martin S. Lindsay, Mary Fletcher, Mary M. Mazziotti, Mary Margaret Cronin, Mary Spears, Matt Mullin, Matthew J. Smith, Maureen Forys, Meaghan Walsh Gerard, Meg Rosenburg, Megan Orpwood-Russell, Meghanne Phillips, Melinda McDonald, Melinda Parsons, Melisande Charles, Meredith Sward, Michael Cerliano, Michael E. Brown, Michael Lang, Michael Martone, Michael Mejia, Michael W. Young, Michelle Carlson, Middlemarch Films, Mo, Moby Stivers, Moran Shoub, Muffie Meyer, Myrna Jackson, Nancy C. Tipton, Nathan Fowler, Nathan Maxwell Cann, Nathaniel Tarn, Nicholas Androulidakis, Nigel Algar, Nikolaj Sømod, Nolan Bennett, Noumena Press, Occulto Magazine, Oficina de Arquitectura, Olga Zilberbourg, Orly Yadin & Bob Summers, Oscar Byrne, Outsiderart, Pardo Fornaciari, Parker Higgins, Patrick R. Cleary, Pattern Research Inc., Patti Gibbons, Paul Harrington, Paula Russell Weiss, Pelle Aardema, Penelope Swan, Pete Mitchell, Peter Brown, Peter Fleck, Peter Fontilus, Philippe Vilon, Phox Pop Magazine, Pia, Picture Panache Framing, Pil Lindgreen K., Pim Bendt, Pirrallos (The), pole/go, Prof. Shorthair, Prof. José Enrique Pons MD, Prue Dixon, Queenscliff Gallery and Workshop, Queer Astrology Project, Rachel Poliquin, Rachel Robbins, Ramiro Pascual, Rayn, Rebecca Clark, Rebecca Resinski, Rechtsanwalt Christian Kramarz LL.M., Renata Tyszczuk, Renato Casella, René Anderson Benitz, Retrogrouch, Richard Owen, Richard Stim, Rob E., Robert Ashton Jarman, Robert M. Peck, Robert S. Petersen, Robert Vowles, Roberto Tejada, Robert P. Stern, Robyn Hugo McIntyre, Ron Sims, Rosemarie Reed, Rufus Pollock, Rusty Winter, Ruthie Dornfeld, Sacramennah, Sander Feinberg, Sandra Huston, Sandro Berra, Sara Mörtsell, Sarah Barton, Sarah E. Gentile, Sarah Louise Bishop, Sarah M. Pickman, Sarah Zar, Sashareen Morgan, Scott Daris, Scott Malthouse, Scout Paget, Self-Publishing Review, Seth Lederer, Shanell Papp, Shannon K. Supple, Shannon Walsh, Sharon, Shyan Y., Sigmund Petersen, Silvia Fernandez, Silvia Sanchez Di Martino, Simone L. Havel, Sonya L. Moore, Stephanie Pierson, Stephen Bohrer, Stephen Stinehour, Stephen Thomas, Steve Booth, Steve Hynds, Stuart Chittenden, Susan DeLand, Susan Kern, Susan Post, Susan Prior, Susan Serna, Susana Sanchez Gonzalez, Susie B., Suzanne Fox, Suzanne Frances Smith, Suzanne Simon, Suzy Byrnes, Sylvia Wu, T-money Candlestix, Tad Kline, Tanja M Cupples Meece, Terrence McDermott, Terri Parsons, Terry

Castle, Terry Harpold, Tessa Hunkin, Thinkfarm Interactive Inc., Thomas E. Thunders, Thomas F. Dillingham, TIME/IMAGE, Tina Sakura, Tod Mesirow (Robot Body Inc.), Tom Fenyll, Tracey Genet, Underfoot Records, Valentin Bakardjiev, Vibhu Mittal, Vicky Loebel, Vince O'Connor, Walter Schoenknecht, Wanderlust Ceramics, WaxMuseums.net, Wellcome Library, Whitworth and I, William B. Holden, William B. Ashworth, Jr., Wilton Gorske, Wolfgang Schlüter, YD Bar-Ness, Yoonmi Nam, Yvette Frock Gottshall, Zachary Barnhart, Zarino Zappia, Zuzka Kurtz.

Learn more about The Friends of The Public Domain Review *and how you can help support the project here: publicdomainreview.org/support.*

Also available from PDR Press:

Lucian's Dialogues of the Gods
PAPERBACK · 152 PAGES · INTRODUCTORY ESSAY & APPENDICES
A sumptuous new edition of Lucian's comic masterpiece, presented in a novel typographic layout. The satirical dialogues — originally penned in the 2nd century AD — eavesdrop on the Ancient Greek gods, presenting us with a sensational peek behind the curtain of life on Mount Olympus.

The Public Domain Review: Selected Essays, Vol. II
PAPERBACK · 195 PAGES · FULL COLOUR · 99 ILLUSTRATIONS
From synesthetic auras and skeletal tableaux to brainwashing machines and truth-revealing diseases; from laughing gas and cocaine-fuelled poems to Byronic vampires and petty pirates — together these fifteen essays chart a wonderfully curious course through the last five hundred years of history.

The Public Domain Review: Selected Essays, Vol. I
PAPERBACK · 346 PAGES · FULL COLOUR · 146 ILLUSTRATIONS
Talking trees, lost Edens, the social life of geometry, imaginary museums, a disgruntled Proust, a frustrated Flaubert… and much much more. Spread across six themed chapters — Animals, Bodies, Words, Worlds, Encounters and Networks — the book includes a stellar line up of contributors, including Jack Zipes, Frank Delaney, Noga Arikha, and Julian Barnes.

www.publicdomainreview.org/pdr-press

Lightning Source UK Ltd.
Milton Keynes UK
UKHW051044250820
368754UK00008B/105